Hormonal Proteins and Peptides

VOLUME XIII

Hormonal Proteins and Peptides

Editor CHOH HAO LI

Laboratory of Molecular Endocrinology
University of California
San Francisco, California

Vol I	1973
Vol II	1973
Vol III	1975
Vol IV	Growth Hormone and Related Proteins 1977
Vol V	Lipotropin and Related Peptides 1977
Vol VI	Thyroid Hormones 1978
Vol VII	Hypothalamic Hormones 1979
Vol VIII	Prolactin 1980
Vol IX	Techniques in Protein Chemistry
Vol X	β-Endorphin
Vol XI	Gonadotropic Hormones
Vol XII	Growth Factors
Vol XIII	Corticotropin (ACTH)

HORMONAL PROTEINS AND PEPTIDES

Edited by CHOH HAO LI

Laboratory of Molecular Endocrinology
University of California
San Francisco, California

VOLUME XIII
Corticotropin (ACTH)

1987

ACADEMIC PRESS, INC.
Harcourt Brace Jovanovich, Publishers
Orlando San Diego New York Austin
Boston London Sydney Tokyo Toronto

COPYRIGHT © 1987 BY ACADEMIC PRESS, INC.
ALL RIGHTS RESERVED.
NO PART OF THIS PUBLICATION MAY BE REPRODUCED OR
TRANSMITTED IN ANY FORM OR BY ANY MEANS, ELECTRONIC
OR MECHANICAL, INCLUDING PHOTOCOPY, RECORDING, OR
ANY INFORMATION STORAGE AND RETRIEVAL SYSTEM, WITHOUT
PERMISSION IN WRITING FROM THE PUBLISHER.

ACADEMIC PRESS, INC.
Orlando, Florida 32887

United Kingdom Edition published by
ACADEMIC PRESS INC. (LONDON) LTD.
24-28 Oval Road, London NW1 7DX

Library of Congress Cataloging in Publication Data

Corticotropin (ACTH).

(Hormonal proteins and peptides; v. 13)
Includes bibliographies and index.
1. ACTH. I. Li, Choh Hao. II. Series [DNLM:
1. Corticotropin. W1 H0626P v. 13 / WK 515 C829]
QP572.P77H67 vol. 13 [QP572.A35] 599.01'927 s 86-32117
ISBN 0-12-447213-3 (alk. paper)   [612'.45]

PRINTED IN THE UNITED STATES OF AMERICA

87 88 89 90   9 8 7 6 5 4 3 2 1

Contents

PREFACE   ix

1   ACTH: Structure–Function Relationship
Ken Inouye and Hideo Otsuka

    I.   Introduction   1
    II.   Structure of ACTH   2
    III.   Structure–Function Relationships   7
    IV.   Conclusions   22
         References   23

2   ACTH Receptors
J. Ramachandran

    I.   Introduction   31
    II.   Detection of ACTH Receptors   32
    III.   Characterization of ACTH Receptors in Rat Adrenocortical Cells   38
    IV.   ACTH Receptors in Human Adrenocortical Cells   48
    V.   ACTH Receptors in Rat Adipocytes   51
    VI.   ACTH Receptors in 3T3-L1 Cells   54
         References   55

3   Biosynthesis of ACTH and Related Peptides
Edward Herbert, Michael Comb, Gary Thomas, Dane Liston, Olivier Civelli, Mitchell Martin, and Neal Birnberg

    I.   Introduction   59
    II.   Structure of POMC Gene and Protein in Different Species   62
    III.   Distribution and Site of Synthesis of POMC-Derived Peptides   64

IV. Regulation of Expression of POMC Genes 66
V. Processing of POMC in the Pituitary and Brain 70
VI. Processing Pathways of POMC in the Anterior and Neurointermediate Lobes of the Pituitary and the Brain 71
VII. Approaches to the Identification of Prohormone Processing Enzymes 74
VIII. Conclusions 80
References 83

4 ACTH and Corticosteroidogenesis

Peter F. Hall

I. Introduction 90
II. Production of Steroids by the Adrenal Cortex 90
III. Stimulation of Steroid Synthesis 97
IV. Site of Action of ACTH in the Steroidogenic Pathway: Cholesterol Transport 98
V. The Role of Cyclic AMP 100
VI. The Role of Protein Synthesis 101
VII. The Role of Phosphorylation 102
VIII. The Role of Ca^{2+} 103
IX. The Possible Role of Protein Kinase C 105
X. The Possible Role of Phospholipids 107
XI. The Roles of Subcellular Components of the Adrenal Cell 110
XII. Synthesis and Conclusions 119
References 121

5 Effect of ACTH and Other Proopiomelanocortin-Derived Peptides on Aldosterone Secretion

Alexander C. Brownie and Robert C. Pedersen

I. Introduction 127
II. Aldosterone Biosynthetic Pathway 128
III. ACTH Action on Aldosterone Biosynthesis 129
IV. Non-ACTH Pituitary Factors Controlling Aldosterone Secretion 133
References 142

6 Behavioral Actions of ACTH and Related Peptides

Curt A. Sandman and Abba J. Kastin

I. Introduction 147
II. Stress 148
III. Opiate-Like Behavioral Effects 151
IV. Learning, Attention, and Memory 152
V. Behavioral Studies in Human Beings 157

VI. Electrophysiological Effects   160
 VII. Developmental Studies: Organizational Influences of Neuropeptides on the Brain   162
VIII. Endogenous Levels   164
  IX. Conclusion   165
       References   166

7 Regulation of ACTH Secretion and Synthesis

Terry D. Reisine and Julius Axelrod

   I. Introduction   173
  II. Multireceptor Release of ACTH   174
 III. Intracellular Mechanisms of ACTH Release   176
  IV. Inhibition of ACTH Release   182
   V. Interactions of Corticotropin Releasing Factors   185
  VI. Desensitization   186
 VII. The Multireceptor Release of ACTH *in Vivo*   189
VIII. Regulation of ACTH Synthesis   191
  IX. Conclusion   193
       References   194

INDEX   197

Preface

The relation of the adrenal to the pituitary has long been suspected from early clinical studies. In 1914, M. Simmonds described his famous case and attributed the clinical disorders to pituitary insufficiency. In 1932, H. Cushing reported cases with hypophyseal basophilic tumors with the hypertrophy of the adrenal cortex. However, the conclusive evidence for the existence of an adrenal-stimulating hormone in the anterior pituitary comes from the classical studies of P. E. Smith in 1930. Smith was the first to demonstrate the extensive atrophy of the adrenal cortex in rats following hypophysectomy and its repair by pituitary implants or injections of pituitary extracts. Thus, the hormone was subsequently designated adrenocorticotropic hormone (ACTH) or corticotropin.

In late 1940, it was demonstrated that a peptide fragment of low molecular weight from ovine pituitary extracts was responsible for the ACTH potency. This led to the isolation of ovine and porcine ACTH with 39 amino acids in 1954. Synthesis of a peptide corresponding to the NH_2-terminal 19 amino acid sequence of ovine ACTH with high biological activity was achieved in late 1960. Total synthesis of porcine ACTH was accomplished in 1963 and human ACTH in 1967.

This volume is devoted to the chemistry and biology of ACTH. The opening chapter on structure–function relationship was written by Inouye and Otsuka who have made significant contributions for over 20 years. This is followed by Ramachandran's chapter on receptors. Early binding studies and problems related to ACTH receptor studies are discussed. Characterization of ACTH receptors in fetal and adult cortical cells is also reviewed.

In 1973, H. D. Moon *et al.* discovered that ACTH and β-lipotropin (β-LPH) are both present in the corticotrophs of sheep pituitary glands. This suggested the possibility that the two hormones may be derived from the same precursor molecule. In 1977, E. Herbert and co-workers showed the existence of proopiomelanocortin (POMC) in mouse pituitary AtT-20 cells. This protein was shown to contain ACTH and β-LPH sequences. In Chapter 3, Herbert reviews current knowledge on the biosynthesis of ACTH primarily in the pituitary but also in extrapituitary tissues.

The adrenal cortex and its secretory products as controlled by ACTH are discussed in the next two chapters. In Chapter 4 Hall considers the role of ACTH in the regulation processes with the production of glucocorticoids. In Chapter 5 Brownie and Pedersen review the mechanism of ACTH action on aldosterone biosynthesis.

The behavior action of ACTH was first observed in 1955 by W. Ferrari *et al.* who reported the stretching and yawning syndrome after the injection of ACTH into the cerebrospinal fluid of dogs. This led to extensive investigations for the last 30 years on the behavior properties of ACTH, β-LPH, and related peptides. In Chapter 6, Sandman and Kastin present a comprehensive analysis of the behavior actions of ACTH, melanotropins, and their peptide fragments.

In 1936, H. Selye proposed the concept of ACTH in the stress response for his general adaptation syndrome. Subsequent studies confirmed the release of ACTH to stressful events under various conditions. Reisine and Axelrod discuss, in the last chapter, regulation of ACTH secretion and synthesis by corticotropin-releasing factor, vasopressin, or catecholamines.

I thank the authors for their contributions and the staff of Academic Press for the production of this volume.

Choh Hao Li

1
ACTH: Structure-Function Relationship

KEN INOUYE AND HIDEO OTSUKA

Shionogi Research Laboratories
Shionogi & Co., Ltd.,
Fukushima-ku, Osaka 553, Japan

I. Introduction

ACTH (adrenocorticotropic hormone, corticotropin) is a peptide hormone which is produced in the anterior lobe of the pituitary and released into the circulation in response to the action of the hypothalamic peptide, corticotropin releasing factor (CRF). The principal action of ACTH is to stimulate steroidogenesis in the adrenal gland to produce glucocorticoids and mineralocorticoids, which act as modulators of a variety of biological responses.

Among the anterior pituitary hormones, ACTH, a linear 39-amino acid peptide, has been studied most extensively with respect to its structure–function relationship. In 1956 it was found that the entire amino acid sequence is not required for the activity of this hormone; the peptide, with only two-thirds of the total sequence, exhibited "full" activity. This important concept is now generally accepted for peptide hormones in general. Together with remarkable advances in the field of peptide chemistry, the synthetic approach to elucidation of structure–activity relationships was carried out most actively in the 1960s and early 1970s (for reviews see Li, 1962; Hofmann, 1962; Lebovitz and Engel, 1964; Schwyzer, 1964; Schröder and Lübke, 1966; Ramachandran and Li, 1967; Ramachandran, 1973; Otsuka and Inouye, 1975; Schwyzer, 1977). As a result we learned that, in the ACTH molecule, the consecutive amino acid residues responsible for receptor recognition [binding site(s)] and those responsible for receptor stimulation [active sites(s)] are arranged in discrete regions. Further advances in structure–function studies must await chemical characterization of the specific receptor as biological counterpart of the hormone. This chapter presents a general survey of the structure–function studies on ACTH with special reference to the steroidogenic activity of synthetic ACTH fragments and analogs. The melanocyte-stimulating hormone (MSH) activity and behavioral effects among the extraadrenal properties of ACTH are also mentioned briefly.

II. Structure of ACTH

A. ACTH

ACTH has been isolated from porcine (White, 1953; Shepherd et al., 1956), ovine (Li et al., 1954, 1955b), bovine (Li and Dixon, 1956), and human (Lee et al., 1959) anterior pituitary glands. More recently, it has also been isolated from dogfish (Lowry et al., 1974), whale (Kawauchi et al., 1978), ostrich (Li et al., 1978), turkey (Chang et al., 1980a), rat (Bennett et al., 1981), and horse (Ng et al., 1981). Structural studies on ovine (Li et al., 1955a), porcine (Howard et al., 1955), bovine (Li et al., 1958), and human (Lee et al., 1961) ACTHs elucidated their 39-amino acid sequences. In 1971, the structures of porcine and human hormones received minor revisions as a result of reexamination of the tryptic fragments (Gráf et al., 1971; Riniker et al., 1972). The structures of ovine and bovine hormones were also revised (Li, 1972; Jöhl et al., 1974). Kawauchi et al. (1978) showed that whale ACTHs isolated from the fin (*Balaenoptera physalus*), sei (*Balaenoptera boleris*), and sperm (*Physeter catodon*) whale pituitaries were all identical with human ACTH in the primary structure. The structures have also been determined for avian ACTHs from ostrich, *Struthio camelus* (Li et al., 1978) and turkey (Chang et al., 1980a), and for an elasmobranchial ACTH from dogfish, *Squalus acanthias* (Lowry et al., 1974). The amino acid sequences of ACTHs from these various species are shown in Table I.

Recent developments in techniques of DNA cloning (Cohen et al., 1973) and nucleotide sequence analysis (Maxam and Gilbert, 1977; Sanger and Coulson, 1978) have made it possible to elucidate the nucleotide sequence coding for the ACTH-β-LPH precursor [pro-opiomelanocortin (POMC)]. Thus, the amino acid sequences of bovine (Nakanishi et al., 1979), human (Chang et al., 1980b), and porcine (Boileau et al., 1984) ACTHs were deduced from the corresponding nucleotide sequences and found to be identical with those determined previously by the peptide sequencing technique. Rat and mouse ACTHs (Table I) were fully characterized by sequence determination of cDNA clones encoding rat (Drouin and Goodman, 1980) and mouse (Uhler and Herbert, 1983; Notake et al.,

Table I—Primary Structure of ACTH

	1	2	3	4	5	6	7	8	9	10	11	12	13	14	15	16	17	18	19	20
Mammals[a-g]	H-	Ser-	Tyr-	Ser-	Met-	Glu-	His-	Phe-	Arg-	Trp-	Gly-	Lys-	Pro-	Val-	Gly-	Lys-	Lys-	Arg-	Arg-	Pro- Val-
Ostrich[h]															Arg					—
Turkey[i]															Arg	Arg	Lys			Ile
Dogfish[j]													Met		Arg					Ile
Salmon I[k]															K/R	K/R	K/R[m]			—
Salmon II[l]													Ile		His					Ile

	21	22	23	24	25	26	27	28	29	30	31	32	33	34	35	36	37	38	39
Human[a] and whale[b]	Lys-	Val-	Tyr-	Pro-	Asn-	Gly-	Ala-	Glu-	Asp-	Glu-	Ser-	Ala-	Glu-	Ala-	Phe-	Pro-	Leu-	Glu-	Phe- OH
Porcine[c]											Leu								
Bovine[d] and ovine[e]													Gln						
Rat[f] and mouse[g]									Asn		Thr	Ser		Gly					
Ostrich[h]						Val		Val	Gln	Glu	Thr	Ser		Gly					
Turkey[i]						Ser	Phe	Val	Asx	Glx	Gln	Ala	Ser	Tyr			Val		
Dogfish[j]			Thr			Val			Glu	Gln	Ser	Val		Asn	Met	Gly	Pro	Leu	
Salmon I[k]				Ala	Ser	Ser	Leu		[n]	Asp	Ser			Gly			Ser	Met	
Salmon II[l]									[o]		Ser							Gln	Ala

[a] Lee et al. (1961); Gräf et al. (1971); Riniker et al. (1972); Chang et al. (1980b). [b] Kawauchi et al. (1978). [c] Howard et al. (1955); Gräf et al. (1971); Riniker et al. (1972); Boileau et al. (1984). [d] Li et al. (1958); Li (1972); Nakanishi et al. (1979). [e] Li et al. (1955a); Li (1972); Jöhl et al. (1974). [f] Scott et al. (1974); Drouin and Goodman (1980); Browne et al. (1981). [g] Uhler and Herbert (1983); Notake et al. (1983). [h] Li et al. (1978). [i] Chang et al. (1980a). [j] Lowry et al. (1974). [k] Kawauchi and Muramoto (1979); Kawauchi et al. (1980). [l] Kawauchi et al. (1981, 1982); Soma et al. (1984).

[m] K/R, not identified, Lys or Arg.
[n] Gly-Gly.
[o] Gly-Thr.

1983) POMCs. Kawauchi *et al.* (1981, 1982) identified two different forms of α-MSH and corticotropin-like intermediate lobe peptide (CLIP) in the pituitary of chum salmon, *Oncorhynchus keta,* and postulated the existence of two putative ACTHs, I and II (Table I). Soma *et al.* (1984) isolated a cDNA clone coding for a protein which had an overall organization similar to those of mammalian POMCs and contained amino acid sequences corresponding to α-MSH II and CLIP II, the less abundant species in the salmon pituitary. The protein containing the predominant species (α-MSH I and CLIP I) has not been found.

Table I shows that structural differences among mammalian ACTH molecules occur at four positions, but there is no difference between human and whale ACTHs, bovine and ovine ACTHs, and rat and mouse ACTHs. Porcine ACTH and bovine/ovine ACTH differ from the human hormone by a single amino acid residue at position 31 (Leu in porcine vs Ser in human) and position 33 (Gln vs Glu), respectively, and rat/mouse ACTH by two amino acid residues at positions 26 (Val vs Gly) and 29 (Asn vs Asp). Serine-31 is known as the phosphorylation site in rat (Browne *et al.*, 1981; Bennett *et al.*, 1981) and human (Bennett *et al.*, 1983) ACTHs and is replaced by leucine in porcine ACTH. Porcine ACTH therefore has no phosphorylation site. Phosphorylation occurs during processing of POMC in the pituitary gland. The phosphorylated and nonphosphorylated forms of human ACTH are equipotent in steroidogenesis in isolated adrenal cells (Bennett *et al.*, 1983).

When compared with mammalian ACTHs, the nonmammalian hormones show much wider variations in structure (Table I). Those variations are mostly located in the carboxyl-terminal region of the molecule. Salmon II sequence has two extra residues, inserted between Gly-29 and Asp-30 and between Gly-34 and Phe-35. In the amino-terminal region, however, the first 12 residues are identical for all species. The next 8 residues in positions 13–20 show some variations, but they are limited to the isosteric replacements of Val by Ile/Met in positions 13 and/or 20 and to the isofunctional replacements of Arg by Lys or Lys by Arg in positions 15–17, except for the putative salmon II sequence in which the residue in position 15 is replaced by His. From these facts, it is apparent that the amino-terminal residues are more important for activity and therefore were better conserved than the carboxyl-terminal residues during the evolution from dogfish to man in order to maintain the biological function of the ACTH molecule.

ACTH is believed to exist in solution without any definite secondary structure. Evidence for some kind of intramolecular interaction in ACTH was presented by Eisinger (1969), followed by reports which showed the appearance of local α-helical structures with increasing concentration of trifluoroethanol in the medium (Löw et al., 1975; Greff et al., 1976). However, Holladay and Puett (1976) confirmed the random conformation of ACTH by a careful study of circular dichroism spectra. Thus, ACTH presents a striking contrast to the insulin molecule which has a rigid three-dimensional structure. In ACTH, the sequential order of amino acid residues is responsible for biological activity, while in insulin the three-dimensional organization of the molecule is thought to be more important. However, it is generally believed that a linear and flexible peptide also possesses some definite structure, an active conformation, when it interacts with the specific receptors. Recently, in the search for such an active conformation of α-MSH [Ac-ACTH-(1–13)-NH$_2$], Sawyer et al. (1982) found that a cyclic analog, [Cys^4, Cys^{10}]-α-MSH, is extremely more active than α-MSH in the frog skin assay for MSH activity and thus have proposed a reverse turn conformation for the α-MSH molecule acting on the MSH receptor.

The total synthesis of porcine ACTH was first accomplished by Schwyzer and Sieber (1963, 1966). Human ACTH was synthesized by Bajusz et al. (1967). However, minor errors were discovered in the previously reported structures of porcine and human ACTHs (Riniker, 1971; Riniker et al., 1972; Gráf et al., 1971) and bovine/ovine ACTH (Li, 1972; Jöhl et al., 1974). Conventional solution methods were used to synthesize the revised sequences of porcine ACTH (Yajima et al., 1976; Inouye et al., 1977), human ACTH (Sieber et al., 1972; Kisfaludy et al., 1972; Nishimura et al., 1975; Koyama et al., 1976b; Inouye et al., 1977), and bovine ACTH (Koyama et al., 1976a), dogfish ACTH (Shimamura et al., 1978), and ostrich ACTH (Yasumura et al., 1982). A solid-phase synthesis of human hormone has also been described by Yamashiro and Li (1973).

B. ACTH-Related Peptides

Several peptides structurally related to ACTH have been characterized in the pituitary of various species. α-Melanocyte-stimulating hormone (α-melanotropin, α-MSH) was isolated from the intermediate lobe of the pituitary and chemically characterized to be a tridecapeptide, Ac-Ser-Tyr-Ser-Met-Glu-His-Phe-Arg-Trp-Gly-Lys-Pro-Val-NH$_2$ (Harris and Lerner, 1957). This amino acid sequence has the first 13 amino acid residues of ACTH. Another melanotropic hormone, β-MSH, was also isolated from mammalian pituitaries. Human β-MSH is a docosapeptide containing the Met-Glu-His-Phe-Arg-Trp-Gly sequence corresponding to positions 4–10 of the ACTH molecule (Harris, 1959; Pickering and Li, 1963). β-MSH was thought to be the principal melanotropic hormone in humans, although its physiological function remained uncertain. Later, β-lipotropin (β-LPH) was isolated from ovine pituitaries (Li, 1964; Li et al., 1965), and when the structure of human β-LPH was elucidated (Cseh et al., 1972), it became apparent that the entire structure of β-MSH was present in the middle of the 91-amino acid sequence of β-LPH. Thus, β-MSH was suggested to be an artifact derived from β-LPH during extraction of the tissue (Scott and Lowry, 1974; Bloomfield et al., 1974). β-LPH attracted much attention when Li and Chung (1976) isolated a 31-amino acid peptide from camel pituitaries. This peptide was found to correspond to the carboxyl-terminal residues 61–91 of ovine β-LPH and was named β-endorphin because of its strong morphine-like activity.

Corticotropin-like intermediate lobe peptide was isolated from the intermediate lobe of rat and pig pituitaries (Scott et al., 1973) and was shown to be identical with positions 18–39 of ACTH. Scott et al. (1973) proposed a possible mechanism for the formation of α-MSH and CLIP from ACTH in the pituitary. The biological function of CLIP is not clear. Porcine CLIP and ACTH (17–38) were shown to stimulate insulin secretion from isolated islet tissue, and this was consistent with the observation that porcine ACTH stimulates insulin release *in vitro* but ACTH (1–24) does not (Beloffe-Chain et al., 1977).

Studies on the biosynthesis of ACTH demonstrated the existence of a large precursor protein to ACTH (Mains et al., 1977; Roberts and Herbert, 1977a,b; Nakanishi et al., 1976, 1977). This protein was also shown to contain β-LPH and β-endorphin in its molecule and was named pro-opiocortin (Rubinstein et al., 1978) or pro-opiomelanocortin (Chrétien et al., 1979). In 1979 Nakanishi et al. reported the complete nucleotide sequence of a cloned cDNA encoding bovine POMC, and they were able to define the precise locations of ACTH and related peptides in the precursor molecule (Nakanishi et al., 1979). Since then human (Chang et al., 1980b), rat (Drouin and Goodman, 1980), mouse (Uhler and Herbert, 1983; Notake et al., 1983), porcine (Boileau et al., 1984), and salmon (Soma et al., 1984) POMCs have been characterized and found to have the same structural organization as that of the bovine POMC molecule.

The steps involved in generating the biologically active peptides from their common precursor have been most extensively studied with mouse and rat POMCs (for reviews see Civelli et al., 1984; Herbert, this volume). In the anterior lobe of the pituitary, POMC is processed to give rise to ACTH and β-LPH under positive regulation by CRF and negative regulation by endogenous glucocorticoids derived from the adrenal cortex. Further processing does not occur in the anterior lobe. In the intermediate lobe of the pituitary, however, the precursor is further processed to cause formation of α-MSH and CLIP from ACTH and formation of β-endorphin and γ-LPH from β-LPH. These processes are not affected by

CRF or glucocorticoids. The marked difference in POMC processing observed between the two lobes of the pituitary represents a typical example of the tissue-specific processing of a common precursor. In the course of processing, POMC and POMC-derived intermediate peptides undergo modifications, such as glycosylation, phosphorylation, acetylation and amidation, in order to become active or to prepare for their secretion.

ACTH and related peptides are found predominantly in the pituitary gland. However, they can also be extracted from the normal placenta and extrapituitary sites of the brain (Krieger *et al.*, 1980). The presence of ACTH immunoreactivity has been demonstrated in the gastrointestinal tract and pancreas of several species (Larsson, 1977, 1978; Orwoll and Kendall, 1980). POMC is probably present in the brain (Yoshimoto *et al.*, 1977), placenta (Liotta *et al.*, 1977), gastrointestinal tract (Hoellt *et al.*, 1978), and a wide variety of tissues of the rat (Saito and Odell, 1983). POMC mRNA has been detected in various regions of the rat brain (Civelli *et al.*, 1982). These observations suggest that POMC is synthesized and processed in various sites in the endocrine and nervous systems, generating ACTH and related peptides.

III. Structure-Function Relationships

A. BIOASSAYS OF ACTH

ACTH possesses a variety of biological functions (Li, 1962) including the ability to stimulate adrenal steroidogenesis and some extraadrenal activities such as the stimulation of melanocytes and the lipolytic action in adipose tissues. The discovery of ACTH-like immunoreactivity in the central nervous system and in the gastrointestinal tract (Larsson, 1978) suggested a broader scope of roles for ACTH and related peptides. The behavioral effects of ACTH-related neuropeptides are particularly noteworthy (for reviews see de Wied, 1969, 1977; van Nispen and Greven, 1982; Kastin, this volume).

In the structure-function studies on ACTH peptides, several bioassay methods have been used to estimate the adrenal-stimulating activity. The classical adrenal ascorbic acid depletion method (Sayers *et al.*, 1948) was useful for following the purification steps of ACTH from natural sources. For some time, this method was also employed for measuring the potencies of synthetic peptides.

An *in vitro* method for estimating corticosterone synthesis in quartered rat adrenal glands was developed by Saffran and Schally (1955). Highly sensitive assay systems for *in vitro* steroidogenesis were developed using the preparation of isolated adrenal cells (Klopenborg *et al.*, 1968; Sayers *et al.*, 1971), which can respond to physiological concentrations of ACTH and are useful for the assay of many samples over a wide range of ACTH concentrations.

Guillemin *et al.* (1958) devised an assay method for *in vivo* steroidogenesis based on elevation of the plasma corticosterone levels in the rat. Lipscomb and Nelson (1962) developed a sensitive *in vivo* method, in which an ACTH preparation was administered to hypophysectomized rats, and the corticosterone concentration in adrenal venous blood was measured. Vernikos-Danellis *et al.* (1966) also demonstrated that adrenal corticosterone levels in hypophysectomized rats are a sensitive measure of ACTH activity.

For measurement of the melanocyte-stimulating activity of ACTH, MSH, and related peptides, the *in vitro* frog skin assay of Shizume *et al.* (1954) has been most commonly employed. The method is based on the dispersion of melanosomes (melanine granules) within dermal melanocytes leading to darkening of the skin.

The behavioral activity of ACTH and related peptides has been assessed by several test systems, including active avoidance behavior, the acquisition or extinction of which is measured using the shuttle box apparatus or the pole-jumping test with rats (for reviews see de Wied, 1977; Beckwith and Sandman, 1978; van Nispen and Greven, 1982).

B. Adrenal-Stimulating Activity

1. *Minimum Structure Essential for Activity*

Table I shows that substantial structural differences among ACTHs from various species occur in the carboxyl (C)-terminal region (positions 26–39, except for the putative salmon sequences) of the ACTH molecules. Therefore, it had been suspected that a considerable portion of the C-terminal sequence of ACTH was not important for the adrenal-stimulating action of the hormone. This was first confirmed by the isolation of a "fully" active peptide, ACTH (1–24), which corresponded to the first 24 amino acid residues of ACTH and was produced by peptic digestion and subsequent limited acid hydrolysis of porcine ACTH (Bell *et al.*, 1956). Further support for the view that the important part for the biological activity is located within the amino (N)-terminal half of the ACTH molecule came from the syntheses of highly active peptides such as ACTH (1–19) (Li *et al.*, 1960), ACTH (1–24) (Kappeler and Schwyzer, 1961), ACTH (1–23) (Hofmann *et al.*, 1961), and ACTH (1–18)-NH_2 (Otsuka *et al.*, 1965c).

The first step in the structure–activity studies of a peptide is usually the search for the shortest fragment that is essential for activity. Table II shows the *in vivo* steroidogenic activity of ACTH (1–24) and related peptides with shorter chain lengths. The activity of ACTH (1–24) may be comparable to that of natural ACTH on a weight basis. Stepwise shortening of ACTH (1–24) from the C-terminal down to ACTH (1–18) causes a gradual decrease in activity. Further removal of the amino acid residues including a part or all of the basic amino acid sequence Lys-Lys-Arg-Arg in positions 15–18 produces remarkable activity losses. It should be noted that ACTH (1–13)-NH$_2$ is only slightly active but still possesses a consistent potency. Nakamura (1972) has shown that [Gly^1]-ACTH (1–14) is only 0.024% active compared to porcine ACTH but exhibits the same maximum level of corticosterone production in the isolated adrenal cells. Therefore, [Gly^1]-ACTH (1–14) and porcine ACTH have the same "intrinsic activity" (Ariëns *et al.*, 1964) but different affinities for adrenal receptors. ACTH (1–10) has been reported to retain activity (Table II) and to exhibit the full intrinsic activity (Schwyzer *et al.*, 1971).

Table III shows the effect of deletion of amino acid residues at the N-terminal of ACTH (1–24) or ACTH (1–23)-NH$_2$. Removal of the first amino acid residue (Ser-1) resulted in a 50% loss of activity (Geiger *et al.*, 1964b). Removal of the first three, four, and five residues produced peptides with residual activities of 15–20, 1, and 0.1%, respectively, and removal of the first six residues including His-6 caused complete loss of activity (Fujino *et al.*, 1971a). These results were later confirmed by the finding that ACTH (7–23)-NH$_2$ and ACTH (7–24) were inactive but ACTH (6–24) stimulated steroidogenesis in the isolated adrenal cell system (Sayers *et al.*, 1974; Fauchére and Petermann, 1978). These findings

Table II—Effect of Chain Length on Steroidogenic Activity (1)

Peptide	Relative *in vivo* activity[a] (%)	Reference
ACTH (1–24)	100	Kappeler and Schwyzer (1961)
ACTH (1–23)	91	Hofmann *et al.* (1961; 1962b)
ACTH (1–19)	47	Li *et al.* (1960, 1961); Ramachandran *et al.* (1965)
ACTH (1–18)	27	Otsuka *et al.* (1965c); Inouye *et al.* (1976b)
ACTH (1–17)	5	Li *et al.* (1962); Ramachandran *et al.* (1965)
ACTH (1–16)	0.1	Hofmann and Yajima (1962)
ACTH (1–13)-NH$_2$	<0.1	Hofmann and Yajima (1961)
ACTH (1–10)	0.002	Ney *et al.* (1964)

[a] Expressed as a percentage of the activity of ACTH (1–24) on a weight basis.

Table III—Effect of Deletion or Addition of Amino Acid Residues at the N-Terminal on *in Vivo* Steroidogenic Activity

Peptide	Relative activity[a] (%)	Reference
ACTH (7–23)-NH$_2$	0	Fujino et al. (1971a)
ACTH (6–24)	0.1	Fujino et al. (1971a)
ACTH (5–23)-NH$_2$	1	Fujino et al. (1971a)
ACTH (4–23)-NH$_2$	15–20	Fujino et al. (1971a)
ACTH (2–23)-NH$_2$	(51)[b]	Geiger et al. (1964b)
ACTH (1–23)-NH$_2$	(100)[b]	Geiger et al. (1964a)
ACTH (1–24)	100	Kappeler and Schwyzer (1961)
Leu-ACTH (1–24)	20	Fujino (1971)
Leu-Leu-ACTH (1–24)	10	Fujino (1971)
Leu-Leu-Leu-ACTH (1–24)	5	Fujino (1971)

[a] Expressed as a percentage of the activity of ACTH (1–24) on a weight basis.
[b] Estimated by the adrenal ascorbic acid depletion assay.

suggest that His-6 is essential for activity, and may constitute part of the active site, while the residues in positions 1–5 are not essential but contribute to the potentiation of the activity. However, one cannot exclude the possibility that the decrease in activity caused by the removal of amino acid residues at the N-terminal may be in part a consequence of transposition of the important α-amino group (see below) from its original place, because a similar decrease has been observed when the N-terminal of ACTH (1–24) is extended by the addition of Leu residues (Fujino, 1971) as shown in Table III.

Schwyzer et al. (1971) reported that ACTH (4–10) was $0.75 \times 10^{-4}\%$ as active as porcine ACTH on a molar basis, and ACTH (5–10) was still active in the isolated adrenal cell system (Sayers et al., 1971). ACTH (4–10) was also shown to cause the same maximum stimulation as ACTH (1–24) in steroidogenesis (Schwyzer et al., 1971). Since ACTH (6–24) elicits weak, but some, steroidogenic activity as mentioned above, it is suggested that the functionally essential structure or the "active center" of the ACTH molecule may reside in the pentapeptide sequence His-Phe-Arg-Trp-Gly (positions 6–10). As mentioned earlier, the same sequence also occurs in the structures of α-MSH and β-MSH. The structure–activity studies on MSH peptides revealed that this common sequence is responsible for the MSH activity (for reviews see Hofmann and Yajima, 1962; Ramachandran and Li, 1967). The synthesis of a tetrapeptide H-His-Phe-Arg-Trp-OH (Otsuka and Inouye, 1964) which possessed MSH activity equivalent to that of the pentapeptide H-His-Phe-Arg-Trp-Gly-OH (Schwyzer and Li, 1958; Hofmann et al., 1958, 1960) established that the tetrapeptide is the minimum structure required for the stimulation of melanocytes. Therefore, it is very likely that the tetrapeptide sequence His-Phe-Arg-Trp is the smallest structure essential for the stimulation of steroidogenesis.

2. Structural Requirements for Full Activity

The N-terminal region of the ACTH molecule contains the essential structural information for steroidogenic activity. However, a short peptide like ACTH (1–10) with low activity needs to combine with a sequence containing at least positions 11–18 in order to acquire a high level of activity (Table II). Competitive antagonism of ACTH (11–24) for both steroidogenesis and cyclic AMP production by ACTH has been demonstrated (Seelig et al., 1971; Seelig and Sayers, 1973). In addition, it was found that a completely inactive fragment ACTH (11–20)-NH_2 binds to an adrenocortical membrane preparation and causes displacement of specifically bound ACTH, and that the ACTH peptides in which the basic amino acid residues in positions 11, 15, 16, 17, and 18 were partially missing or blocked were less effective competitors than ACTH (11–20)-NH_2 (Hofmann et al., 1970b; Finn et al., 1972). These observations clearly indicate that the sequence 11–18 constitutes the specific binding site of the ACTH molecule.

Table IV shows the *in vivo* steroidogenic activity of ACTH peptides with different chain lengths from 17 to 39, where the activities were compared both on a weight basis and on a molar basis. The activity of ACTH (1–24) may be comparable to that of natural ACTH (1–39) on a weight basis, but it is estimated to be about 50% on a molar basis. Ney et al. (1964) assayed a variety of synthetic ACTH peptides, obtained from different laboratories, under identical conditions and found that the molar activity continued to increase as the peptide chain grew longer up to 39. Szporny et al. (1969), who assayed ACTH peptides synthesized by Bajusz et al. (1967, 1968), also found that ACTH (1–39) was more active than ACTH (1–32), ACTH (1–28), and ACTH (1–24). These findings, as well as the data shown in Table IV, clearly indicate that the full-length ACTH is more active than any of the shorter peptides.

Table IV also shows the effect of amidation at the C-terminal of ACTH peptides. ACTH (1–17)-NH_2, ACTH (1–18)-NH_2, and ACTH (1–19)-NH_2, in which the C-terminal α-carboxyl group is blocked in the amide form, are six, four, and three times more active than the corresponding peptide acids, respectively, and the last two peptide amides are almost as active as ACTH (1–39) on a weight basis. In addition, it has been shown that ACTH (1–18)-NH_2 is inactivated more slowly than ACTH (1–18) in human plasma (Imura et al., 1967). The remarkable potentiation associ-

Table IV—Effect of Chain Length on Steroidogenic Activity (2)

Peptide	Activity in vivo		Reference
	Units/mg	Units/µmol	
ACTH (1–17)	5 (4)[a]	11 (2)[a]	Ramachandran et al. (1965)
ACTH (1–17)-NH$_2$	33 (28)	69 (13)	Ramachandran et al. (1965)
ACTH (1–18)	27 (23)	61 (11)	Inouye et al. (1976b)
ACTH (1–18)-NH$_2$	97 (81)	218 (40)	Inouye et al. (1976b)
ACTH (1–19)	47 (39)	110 (20)	Ramachandran et al. (1965)
ACTH (1–19)-NH$_2$	147(123)	345 (63)	Ramachandran et al. (1965)
ACTH (1–23)	91 (76)	258 (47)	Hofmann et al. (1962b)
ACTH (1–23)-NH$_2$	80 (67)	230 (42)	Fujino et al. (1970a)
ACTH (1–24)	100 (83)	293 (54)	Kappeler and Schwyzer (1961)
ACTH (1–26)	160(133)	497 (91)	Inouye et al. (1976c)
ACTH (1–27)	155(129)	492 (90)	Inouye et al. (1976c)
ACTH (1–28)[b]	120(100)	397 (73)	Szporny et al. (1969)
ACTH (1–32)[b]	131(109)	486 (89)	Szporny et al. (1969)
ACTH (1–39), human	120(100)[c]	545(100)	c

[a] The relative activity expressed as a percentage of the activity of synthetic human ACTH is given in parentheses.

[b] The sequence 25–30 is Asp-Ala-Gly-Glu-Asp-Gln and differs from the corrected sequence Asn-Gly-Ala-Glu-Asp-Glu (Gráf et al., 1971; Riniker et al., 1972). There is no significant activity difference between human ACTH and human [Asp^{25}, Ala^{26}, Gly^{27}, Gln^{30}]-ACTH (Sieber et al., 1972).

[c] The value is an average of reported values (99–140 units/mg) for synthetic human ACTH (Sieber et al., 1972; Kisfaludy et al., 1972; Nishimura et al., 1975; Koyama et al., 1976b; Inouye et al., 1977).

ated with the C-terminal amidation is therefore likely to be a consequence of the increased resistance of the ACTH peptides toward the action of carboxypeptidase in the biological systems. However, this effect of amidation becomes less remarkable as the peptide chain is extended at the C-terminal, ACTH (1–23)-NH$_2$ being no longer more active than ACTH (1–23) in in vivo steroidogenesis (Table IV). An in vitro experiment has shown that ACTH (1–18), ACTH (1–19), ACTH (1–18)-NH$_2$, and ACTH (1–26) are inactivated in human serum with half-lives of approximately 0.5, 1, 1.5, and 2 hr, respectively, while under the same conditions the full-length ACTH remains active over a period of 4 hr (Imura et al., 1967).

All these observations strongly suggest that the C-terminal half of the ACTH molecule may not be essential for adrenal stimulation but plays an important role in protecting the functional parts of the molecule from destruction by degradative enzymes, especially during transport to the receptor site of the target tissue.

3. Effect of N-Terminal Substitution

Short chain peptides such as ACTH (1–18)-NH_2 or ACTH (1–24) possess a high level of steroidogenic activity, but they are apparently short-acting compared to native ACTH. Much synthetic effort has been devoted to the development of short-chain peptides which have long-acting properties and seem to be of clinical importance. The considerable inactivation of porcine ACTH by the action of leucine aminopeptidase (White, 1955) suggested the involvement of aminopeptidase(s) in the inactivation of ACTH in tissues. This led to attempts to increase the resistance toward aminopeptidases.

In the early stage of studies on ACTH, chemical modification of the N-terminal serine residue by either selective acetylation (Waller and Dixon, 1960) or periodate oxidation (Dixon, 1956) yielded products with little or no adrenal-stimulating activity. It was also demonstrated that the 1-glycolic acid ($HOCH_2CO$-) analog of porcine ACTH was devoid of activity (Dixon, 1962), while the corresponding 1-glycine (NH_2CH_2CO-) analog was as active as the intact hormone (Dixon and Weitkamp, 1962). These results, indicating that the amino group but not the hydroxymethyl group of the N-terminal serine residue was important for activity, were confirmed by the synthesis of [Gly^1]-ACTH (1–18)-NH_2 (Otsuka et al., 1970) and [Gly^1]-ACTH (1–23)-NH_2 (Geiger et al., 1969). These and some of the synthetic octadecapeptides and tetracosapeptides in which the N-terminal residue is replaced by other amino acids are shown in Table V.

In view of the importance of the N-terminal α-amino group of ACTH, the modification to increase the resistance to aminopeptidase action had to be done without altering the amino function. To accomplish this, Boissonnas et al. (1966) synthesized [D-Ser^1, Nle^4, Val^{25}]-ACTH (1–25)-NH_2 containing D-serine in place of serine at the N-terminal. Kappeler et al. (1967) introduced D-serine or D-alanine into position 1 of ACTH (1–24) (Table V). These 1-D-amino acid analogs were found to be 3 to 10 times more active than the corresponding all-L peptides. Inouye et al. (1970b) examined the effect of substitution at the N-terminal by preparing [βAla^1]-ACTH (1–18)-NH_2 and [Aib^1]-ACTH (1–18)-NH_2. Comparison of the biological activities of these analogs with those of ACTH (1–18)-NH_2 revealed that substitution of the N-terminal with β-alanine or α-aminoisobutyric acid enhances the activity of the peptide, especially in the case of the Aib^1-analog which was 10 times more activity than the Gly^1-analog (Table V). Measurements of the half-lives of these octadecapeptides and native ACTH showed that the substitution with β-alanine or α-aminoisobutyric acid increases the stability of the peptide in tissues but does not significantly increase the stability in plasma, and that the native ACTH

Table V—Effect of N-Terminal Substitution on *in Vivo* Steroidogenic Activity

Peptide	Relative activity (%)	Reference
[X^1]-ACTH (1–24)		
X = Ser	100	Kappeler and Schwyzer (1961)
X = Gly[a]	104	Geiger *et al.* (1969)
X = Pro	50–65	Fujino *et al.* (1970b)
X = Lys	30–50	Fujino *et al.* (1970b)
X = D-Ala	300	Kappeler *et al.* (1967)
X = D-Ser	300–1000	Kappeler *et al.* (1967)
X = βAla	100–160	Fujino *et al.* (1970b)
X = γAbu[b]	100–110	Fujino *et al.* (1970b)
X = Sar[b]	100–160	Fujino *et al.* (1970b)
[X^1]-ACTH (1–18)-NH$_2$		
X = Ser	100	Otsuka *et al.* (1965c); Inouye *et al.* (1976b)
X = Gly	130	Otsuka *et al.* (1970)
X = βAla	290	Inouye *et al.* (1970a)
X = Aib[b]	1240	Inouye *et al.* (1970b)

[a] [Gly^1]-ACTH (1–23)-NH$_2$.
[b] Abbreviations: γAbu, γ-aminobutyric acid, NH$_2$(CH$_2$)$_3$COOH; Sar, sarcosine, CH$_3$NHCH$_2$COOH; Aib, α-aminoisobutyric acid, NH$_2$C(CH$_3$)$_2$COOH.

exhibited longer duration of action than the βAla^1- and Aib^1-octadecapeptides in plasma (Inouye *et al.*, 1970b; Tanaka, 1971). The lower activity of [βAla^1]-ACTH (1–18)-NH$_2$ compared to [Aib^1]-ACTH (1–18)-NH$_2$ was interpreted as indicating the importance of the distance between the amino group at the N-terminal and the active center of the peptide, although the difference between α- and β-amino groups in basicity might also be of some significance.

4. *Effect of Substitution in Positions 6–9 and Their Vicinities*

The sequence His-Phe-Arg-Trp (positions 6–9) not only constitutes the active site of the ACTH molecule but also occurs in common in the molecules of ACTH, MSH, and LPH. The tetrapeptide corresponding to this sequence has been known to exhibit significant MSH and lipolytic activities (Otsuka and Inouye, 1964). In addition, an ACTH peptide, H-Met-Glu-His-Phe-Arg-Trp-Gly-OH (positions 4–10), has been shown to be as active as native ACTH with respect to the behavioral effects in the rat (for a review see van Nispen and Greven, 1982). In view of these facts, it should be interesting to examine the effect of substitution in this region in terms of the adrenal as well as extraadrenal activities. The data shown in Table VI are those reported for synthetic ACTH analogs with chain lengths from (1–17)-NH_2 to (1–24).

The side-chain imidazole of His-6 may be replaceable by pyrazolyl, but the presence of a dissociable group at this position is probably important for steroidogenic activity since the [Phe^6]-analog was only slightly active. It is noteworthy that the [Pal^6]-analog (for abbreviations see footnote in Table VI) showed a greater decrease in MSH activity than in steroidogenic activity (Table VI)—this case, in which amino acid substitution affected the steroidogenic activity less than the MSH activity, is an unusual one.

The phenylalanine in position 7 is partially replaceable by leucine. The replacement by D-phenylalanine caused a great decrease in steroidogenesis while it brought about a dramatic increase in MSH activity (Table VI). A [D-Phe^7]-analog and the parent peptide were found to stimulate steroidogenesis to the same maximal level in isolated adrenal cells, although the D-Phe-peptide required a much higher concentration (Nakamura, 1972). The L-configuration may be required at position 7 for tighter binding with adrenal receptors. In ACTH (4–10) or [Lys^8]-ACTH (4–9), the substitution of each of the constituent amino acid residues by the corresponding D-isomer usually causes potentiation of the acquisition or delay of extinction of a conditioned avoidance behavior in the rat. However, this was not the case with Phe-7, in which the replacement by D-Phe brought about acceleration instead of delay of extinction of the avoidance behavior (Bohus and de Wied, 1966).

Replacement of the guanidino group of Arg-8 by an amino group to give an [Orn^8]-analog (Orn = ornithine) results in a great loss of both steroidogenic and MSH activities (Table VI). Lengthening of the side-chain of Arg-8 by one methylene unit gives a [Har^8]-analog (Har = homoarginine). The steroidogenic activity of [Har^8]-ACTH (1–24), for example, is reduced to 25% of that of the parent peptide but still higher than that of the [Lys^8]-analog (Tesser et al., 1973a). [Nar^8]-ACTH (1–24) (Nar = norarginine with a side-chain shorter than that of arginine by one methylene unit) seems to be less active than the corresponding [Har^8]-analog, while [Nar^8, Lys^{17}, Lys^{18}]-ACTH (1–18)-NH_2 is reported to be nearly as active as [Lys^{17}, Lys^{18}]-ACTH (1–18)-NH_2 in in vivo steroidogenesis (van Nispen et al., 1977b). These observations indicate that the presence of a guanidino group in position 8 is of critical importance, and the length of the side chain seems to be of less importance for steroidogenic activity and probably for MSH activity as well. The substitution of Arg-8 by D-arginine led to substantial loss of steroidogenic activity (Table VI).

The behavioral effect of ACTH (4–10) in the rat is not affected by the substitution of Arg-8 by lysine. However, the substitution by D-arginine,

Table VI—Effect of Substitution in Positions 6-11 on the Biological Activities of ACTH

Substitution[a]						Steroidogenic activity (%)	MSH activity (%)	Reference
6	7	8	9	10	11			
-His—	-Phe—	-Arg—	-Trp—	-Gly—	-Lys-	100	100	
Pal	—	—	—	—	—	72	10	Hofmann et al. (1970a)
Phe	—	—	—	—	—	1.4	3	Blake and Li (1972)
—	Leu	—	—	—	—	20	—	Fujino et al. (1971b)
—	D-Phe	—	—	—	—	5	680	Inouye et al. (1971)
—	—	Lys	—	—	—	7.6	5	Chung and Li (1967); Tesser et al. (1973a)
—	—	Orn	—	—	—	3.6	10	Tesser et al. (1973a); Tesser and Rittel (1969)
—	—	Har	—	—	—	30–100	—	Tesser et al. (1973a,b)
—	—	Nar	—	—	—	3–10	8–27	van Nispen et al. (1977a,b)
—	—	D-Arg	—	—	—	0.002	—	Löw et al. (1980)
—	—	—	Phe	—	—	2.4	9–13	Hofmann et al. (1970a)
—	—	—	Ile	—	—	0	—	Kumar (1975)
—	—	—	Trp(Nps)	—	—	1.5	340	Canova-Davis and Ramachandran (1976)
—	—	—	Trp(For)	—	—	18	48	Blake and Li (1975)
—	—	—	MeTrp	—	—	0.05–1	—	Fujino et al. (1971b)
—	—	—	Pmp	—	—	0.15	21–42	van Nispen and Tesser (1975); van Nispen et al. (1977a)
—	—	—	Nal	—	—	7.4	40	Blake and Li (1975)
—	—	—	—	—	Arg	100	—	Geiger and Schröder (1973)
—	—	—	—	—	Nle	0.7	—	Geiger and Schröder (1973)
—	—	—	—	—	Gly	0.1	1.3	Shin et al. (1971); Geiger and Schröder (1973)

[a] Abbreviations: Pal, L-β-(pyrazolyl-3)alanine; Har, L-homoarginine; Nar, L-norarginine; Trp(Nps), N^{in}-(2-nitrophenylsulfenyl)-L-tryptophan; Trp(For), N^{in}-formyl-L-tryptophan; MeTrp, N^a-methyl-L-tryptophan; Pmp, L-pentamethylphenylalanine; Nal, L-β-(naphthyl-1)alanine.

and that by D-lysine, led to the production of peptides which were 3 and 10–30 times more active than the reference compound ACTH (4–10), respectively (van Nispen and Greven, 1982).

The effects of substitution at position 9 have been most extensively studied. The replacement of Trp-9 by phenylalanine in [Gln^5]-(1–20)-NH$_2$ caused a marked decrease but not a total loss in both *in vivo* steroidogenic and MSH activities (Hofmann *et al.*, 1970a). In view of the fact that the [Phe^9]-analog still possessed a residual 2 to 3% activity, the indole group of Trp-9 seemed not to be functionally involved in steroidogenesis but to contribute to the binding of the hormone to its receptor. The replacement or modification of Trp-9 to produce the [Pmp^9]-, [Nal^9]-, [Trp(Nps)^9]-, and [Trp(For)^9]-analogs of ACTH peptides (Table VI) greatly diminishes the steroidogenic activity but not the MSH activity. In the case of [Trp(Nps)^9]-ACTH (1–39), the steroidogenic activity was only 1.5% that of ACTH, while the MSH activity was increased even more than 3-fold (Ramachandran, 1970, 1973). This adverse effect has also been observed with the [D-Phe^7]-analog as mentioned above and represents clear separation of the two activities. The substitution of Trp-9 by N^α-methyltryptophan also produces analogs with greatly reduced steroidogenic activity, indicating that the integrity of the indole group is not sufficient for the activity. The introduction of the N^α-methyl group may cause a conformational change in the vicinity of Trp-9 which interferes with a close fit of the indole and its neighboring groups to the ACTH receptor. All of these analogs having an aromatic substituent at position 9 possess low but definite steroidogenic activity. Most of them are able to stimulate steroidogenesis to the same level as ACTH, even though much higher concentrations of the modified peptides are required. [Trp(Nps)^9]-ACTH (1–24) and [Trp(Nps)^9]-ACTH (1–39) are reported to be partial agonists, their maximal level of steroidogenesis in isolated adrenal cells being lower than that of the agonists ACTH (1–24) and ACTH (Seelig *et al.*, 1972). However, [Ile^9]-ACTH (1–24), in which Trp-9 was substituted by an aliphatic amino acid isoleucine, failed to stimulate corticosterone and cyclic AMP production (Kumar, 1975).

[Trp(Nps)^9]-ACTH (1–24) and [Trp(Nps)^9]-ACTH (1–39) were shown to antagonize the steroidogenic action of ACTH (1–24) and ACTH in isolated adrenal cells (Seelig *et al.*, 1972). It has also been reported that [Trp(Nps)^9]-ACTH (1–39) inhibits in a competitive manner the ACTH-induced cyclic AMP production but not steroid synthesis (Moyle *et al.*, 1973). [Ile^9]-ACTH (1–24) was shown to inhibit both cyclic AMP production and corticosterone production induced by ACTH (1–24) in isolated adrenal cortex cells (Kumar, 1975). Hofmann *et al.* (1974) carried out direct binding studies on several ACTH analogs using adrenal plasma membrane preparations and showed that [Gln^5, Phe^9]-ACTH (1–20)-NH$_2$ and [MeTrp^9]-ACTH (1–24) failed to stimulate the adenylate cyclase system but exhibited high affinity for the ACTH receptor(s) and strongly antagonized the action of ACTH (1–24). These observations demonstrate that a low adrenal stimulating potency does not directly indicate reduced affinity for the ACTH receptors and emphasize the importance of Trp-9 in the process involved in excitation of the receptors. [Lys^8, Phe^9]-ACTH (1–19) has been reported not to inhibit the steroidogenesis induced by ACTH (1–19) (Blake and Li, 1975), although [Lys^8]-ACTH (1–24) is an antagonist of ACTH (1–24) (Hofmann et al., 1974).

The behavioral effect of ACTH (4–10) in the rat is either not affected or somewhat increased by the substitution of Trp-9 by tyrosine or phenylalanine, respectively, as estimated by the pole-jumping test (van Nispen and Greven, 1982).

Table VI also shows that Lys-11 is fully replaceable by arginine and may be replaceable by ornithine (Tesser and Buis, 1971), but removal of the ε-amino group of Lys-11 to form a [Nle^{11}]-analog results in a marked decrease in steroidogenic activity. The total removal of the side-chain of Lys-11 produces a [Gly^{11}]-analog which has almost no steroidogenic and MSH activities. These observations clearly indicate a high degree of contribution of Lys-11 as a part of the strong binding site (positions 11–18) of the ACTH molecule. This amino acid residue must be basic regardless of the nature of its basicity and represents a marked contrast to Arg-8, which is partially replaceable by isofunctional homoarginine or norarginine but virtually nonreplaceable by lysine or ornithine (Table VI).

Glycine in position 10 may be regarded as a hinge connecting the active site (positions 6–9) and the binding site (11–18) of the ACTH molecule. Inouye et al. (1978) studied the effect of substitution at position 10 by synthesizing the [Ala^{10}]-, [D-Ala^{10}]-, [Aib^{10}]-, and [βAla^{10}]-analogs of ACTH (1–18)-NH$_2$. Table VII shows that all of these analogs are less active than the parent peptide in both the in vivo steroidogenic and the MSH activities. The MSH activity varied in parallel with the steroidogenic activity. The Ala^{10}-analog possesses a reduced but substantial degree of activity, whereas the D-Ala^{10}-analog shows only 1% or less of the activity of the parent peptide with glycine in position 10. Introduction of an additional methyl group to the α-carbon of the position 10 residue of the Ala^{10}- or D-Ala^{10}-analog produces the Aib^{10}-analog, which is virtually inactive. Lengthening of the peptide chain by one methylene unit at position 10 produces the inactive βAla^{10}-analog. Comparison of circular dichroism spectra of [X^{10}]-ACTH (1–18)-NH$_2$ revealed that there is no significant conformational difference between the active peptides (X = Gly or Ala) and the inactive peptides (X = D-Ala, Aib, or βAla) in solution. It

Table VII—Biological Activities of $[X^{10}]$-ACTH (1–18)-NH_2 in Comparison with ACTH (1–18)-NH_2 (X = Gly): Effect of Substitution at Position 10[a]

X	Gly	Ala	D-Ala	Aib	βAla
Formula	$NH-\underset{H}{\overset{H}{C}}-CO$	$NH-\underset{H}{\overset{CH_3}{C}}-CO$	$NH-\underset{CH_3}{\overset{H}{C}}-CO$	$NH-\underset{CH_3}{\overset{CH_3}{C}}-CO$	$NH-\underset{H}{\overset{H}{C}}-\underset{H}{\overset{H}{C}}-CO$
In vivo steroidogenic activity[b]	100	35	1.1	0.17	0.13
MSH activity[b]	100	48	0.39	0.15	0.17

[a] After Inouye et al. (1978).
[b] Percentage.

has also been shown that the inactive peptides do not antagonize the MSH activity of ACTH (1–18)-NH$_2$ and α-MSH (Inouye et al., 1978).

The flexibility of the peptide backbone must be promoted by the presence of a glycine residue, but it may be reduced when the α-hydrogen atoms of the glycine are substituted by one or two methyl groups as in the ACTH analogs with Ala, D-Ala, or Aib at position 10. The reduced flexibility will make it difficult for the peptide to take the biologically active conformation when it interacts with specific receptors. When the glycine residue is substituted by β-alanine as in [βAla^{10}]-ACTH (1–18)-NH$_2$, an extra methylene unit which is inserted into the peptide backbone causes a so-called frame shift and alters the steric relations of the side-chain groups, probably leading to a significant change in the conformation of the peptide.

5. Effect of Substitution in Other Positions

ACTH is almost completely inactivated by mild oxidation with hydrogen peroxide, but reduction with cysteine or thioglycolic acid can restore the activity (Dedman et al., 1955). This phenomenon is apparently associated with the methionine–methionine sulfoxide interconversion (Dedman et al., 1961). The methionine residue in position 4 thus seemed to be important for activity. However, this amino acid residue was later found to be 40% replaceable by α-aminobutyric acid (Hofmann et al., 1963, 1964) and fully replaceable by an isosteric amino acid norleucine (Doepfner, 1969). Therefore, the inactivation of ACTH caused by the conversion of a methionine into methionine sulfoxide may be a consequence of a remarkable increase in polarity in the vicinity of Met-4. This finding agrees with the view that Met-4 constitutes a part of the hydrophobic binding site of the ACTH molecule. [Phe^2, Nle^4]-ACTH (1–38) was synthesized and found to be equipotent with human ACTH (Buckley et al., 1981a). This analog was iodinated to give [^{125}I-labeled Tyr^{23}, Phe^2, Nle^4]-ACTH (1–38) which was fully active in stimulating steroidogenesis and utilized in receptor-binding studies (Buckley et al., 1981b). By using this analog, the iodination at Tyr-2 and the oxidation of Met-4, which cause marked reduction of activity, were eliminated.

ACTH contains three arginine residues in positions 8, 17, and 18. Arginine in position 8 occurring in the active site cannot be substituted by lysine, as discussed above, while Arg-17 and Arg-18, which constitute a part of the binding site of the ACTH molecule, can be replaced by lysines (Desaulles et al., 1969) or ornithines (Tesser and Schwyzer, 1966). Otsuka et al. (1965a,b) synthesized analogs in which one or both of the lysines in positions 15 and 16 were substituted by arginines, but these peptides were found to be less active than the parent peptides. The lysine peptides are usually easier to handle than the corresponding arginine peptides when the synthesis is done by the conventional solution methods (for peptide synthesis see Bodanszky et al., 1976). Such peptides include [D-Ser^1, Lys^{17}, Lys^{18}]-ACTH (1–18)-NH$_2$ (Riniker and Rittel, 1970) and [Aib^1, Lys^{17}, Lys^{18}, Lys^{19}]-ACTH (1–19)-NH$_2$ (Inouye et al., 1976a) which were synthesized with the aim of possible clinical application.

C. Melanocyte–Stimulating Activity

The structural similarities between ACTHs and MSHs have already been mentioned. Among the extraadrenal effects of ACTH, the melanocyte-stimulating activity has been well investigated and discussed as a function of structure (Hofmann and Yajima, 1962; Ramachandran and Li, 1967; Schwyzer, 1977; Hruby et al., 1984).

The sequence of the first 13 amino acid residues of ACTH is identical with that of α-MSH. Therefore, it is not surprising that ACTH exhibits melanocyte stimulation, although its potency is only 1% that of α-MSH. Table VIII shows the MSH activity of selected members of ACTH peptides, including ACTH, α-MSH and their synthetic fragments. The smallest active fragment that is common to ACTH, α-MSH, and β-MSH is the tetrapeptide ACTH (6–9); the His-Phe-Arg-Trp sequence is the active site

Table VIII—Melanocyte-Stimulating Activity of ACTH Peptides

Peptide	MSH activity *in vitro*		Reference
	Units/g	Units/mmol	
[Gln^5]-ACTH (1–8)	0	0	Hofmann et al. (1960)
ACTH (6–9)	3.6×10^4	2.3×10^4	Otsuka and Inouye (1964)
ACTH (6–10)	3.1×10^4	2.2×10^4	Schwyzer and Li (1958)
ACTH (11–13)-NH$_2$	—	3×10^4	Eberle and Schwyzer (1975)
ACTH (7–13)-NH$_2$	4.0×10^5	3.6×10^5	Hofmann and Yajima (1962)
Ac-ACTH (7–13)-NH$_2$	—	5×10^6	Eberle and Schwyzer (1975)
ACTH (6–13)-NH$_2$	8.0×10^6	8.2×10^6	Hofmann and Yajima (1961)
Ac-ACTH (5–13)-NH$_2$	—	5×10^8	Eberle and Schwyzer (1975)
ACTH (1–13)-NH$_2$	1.9×10^9	3.1×10^9	Hofmann and Yajima (1961)
Ac-ACTH (1–13)-NH$_2$ (α-MSH)	2.0×10^{10}	3.4×10^{10}	Lee and Lerner (1956)
	1.5×10^{10}	2.5×10^{10}	Guttmann and Boissonnas (1959); Schwyzer et al. (1963)
ACTH (1–16)	3.7×10^8	7.2×10^8	Hofmann et al. (1962a)
ACTH (1–24)	1.2×10^8	3.5×10^8	Schwyzer and Kappeler (1963)
Ac-ACTH (1–24)	1.0×10^9	3.0×10^9	Schwyzer and Kappeler (1963)
ACTH (1–39)	1.2×10^8	5.4×10^8	Ney et al. (1964)
Ac-ACTH (1–39)	7.2×10^8	3.3×10^9	Waller and Dixon (1960)

for the MSH as well as steroidogenic activities of ACTH. Eberle and Schwyzer (1975) found that tripeptide derivatives, H-Lys-Pro-Val-NH$_2$ and Ac-Lys-Pro-Val-NH$_2$, corresponding to the 11–13 sequence of ACTH possessed specific MSH activity with potencies comparable to that of ACTH (6–9). This tripeptide sequence not occurring in β-MSH was regarded as the second active site of α-MSH. Combination of the two active fragments ACTH (6–9) and ACTH (11–13)-NH$_2$ greatly enhances the activity as seen in ACTH (6–13)-NH$_2$ (Table VIII). Further addition of an inactive fragment, ACTH (1–5), causes further remarkable enhancement of activity leading to the production of Ac-ACTH (1–13)-NH$_2$ (α-MSH, 3 to 4 × 10^{10} units/mmol). Further elongation of the peptide chain at the C-terminal reduces activity to 10^8 units/mmol, where it remains fairly constant for the peptides from ACTH (1–16) to ACTH (1–39).

Acetylation of the N-terminal amino group of ACTH greatly reduces the steroidogenic activity, but the MSH activity of the hormone is increased 6- to 10-fold (Waller and Dixon, 1960). The same effect of acetylation can be observed with peptides of shorter chain lengths (Table VIII). Dissociation of the steroidogenic and MSH activities were similarly observed with substitution of D-phenylalanine for Phe-7 and that of Trp(Nps) for Trp-9 as shown in Table VI. For example, the steroidogenic activity of [βAla^1, Orn^{15}]-ACTH (1–18)-NH$_2$ was mostly lost when Phe-7 was replaced by D-phenylalanine to give [βAla^1, D-Phe^7, Orn^{15}]-ACTH (1–18)-NH$_2$, while the MSH activity was remarkably increased to a level even higher than that of α-MSH and its action was greatly protracted (Inouye et al., 1971). On the other hand, replacement of Ser-1 of ACTH peptides by a D-amino acid, β-alanine, or α-aminoisobutyric acid (see Table V) increases both steroidogenic and MSH activities several-fold (Kappeler et al., 1967; Inouye et al., 1970a,b), and the oxidation of Met-4 to methionine sulfoxide leads to substantial inactivation of both steroidogenic and MSH activities (Dedman et al., 1961). The structure–activity studies on ACTH have shown that the decrease in steroidogenic activity due to structural alteration is not always accompanied by a decrease in MSH activity, but an increase in steroidogenic activity is usually accompanied by a greater increase in MSH activity.

IV. Conclusions

The structure–function relationships discussed in this review, mainly in regard to the action of ACTH on the adrenal gland, lead to the following conclusions.

ACTH is a linear peptide and its primary structure, the sequential order of amino acid residues, is responsible for its biological activities. Considerable evidence suggests that the amino acid sequence His-Phe-Arg-Trp (positions 6–9) constitutes the active site of ACTH, being responsible for the stimulation of specific receptors. The amino acid residues in this portion have strict structural requirements. The arginine residue in position 8, for example, can only be substituted by an isofunctional amino acid such as homoarginine and represents a striking contrast to the other basic amino acid residues occurring in positions 11 and 15–18 which permit exchange between arginine and lysine residues.

The sequence Lys-Pro-Val-Gly-Lys-Lys-Arg-Arg (positions 11–18) is not essential for activity but appears to play, as a major binding site, an important role in recognizing the specific adrenal receptor. The N-terminal sequence Ser-Tyr-Ser-Met-Glu (positions 1–5) may act as a secondary binding site of ACTH. This pentapeptide sequence seems to be less important than the two portions mentioned above, but the α-amino group of Ser-1 is especially important for the steroidogenic activity. Steroidogenesis is extremely lowered when this amino group is blocked, probably as a consequence of alteration in the electrostatic nature or conformation of the ACTH molecule, while the MSH activity is increased several-fold, offering some evidence for clear dissociation of the two activities.

The C-terminal half of the ACTH molecule contains species-specific variations. This portion may not be involved in the interaction with adrenal cells or melanocytes but may be important for protecting the hormone from degradation in the physiological environment. This portion may also be released itself as a peptide like CLIP, the function of which is not yet understood.

Further progress in the structure–function studies of ACTH awaits the isolation and characterization of specific receptor molecules and chemical and biochemical studies on the interaction between hormone and receptor molecules. Such studies should lead to better understanding of the mechanisms of action of ACTH and other peptide hormones.

References

Ariëns, E. J., Simonis, M., and van Rossum, J. M. (1964). *In* "Molecular Pharmacology" (E. J. Ariëns, ed.), Vol. 1, pp. 136–148. Academic Press, New York.
Bajusz, S., Medzihradszky, K., Paulay, Z., and Lang, Zs. (1967). *Acta Chim. Hung.* **52**, 335–341.
Bajusz, S., Paulay, Z., Láng, Zs., Medzihradszky, K., Kisfaludy, L., and Löw, M. (1968). *Proc. Eur. Pept. Symp., 9th, Orsay* pp. 237–242.
Beckwith, B. E., and Sandman, C. A. (1978). *Neurosci. Biobehav. Rev.* **2**, 311–338.

Bell, P. H., Howard, K. S., Shepherd, R. G., Finn, B. M., and Meisenhelder, J. H. (1956). *J. Am. Chem. Soc.* **78,** 5059–5066.
Beloffe-Chain, A., Edwardson, J. A., and Hawthorn, J. (1977). *J. Endocrinol.* **73,** 28–29P.
Bennett, H. P. J., Browne, C. A., and Solomon, S. (1981). *Biochemistry* **20,** 4530–4538.
Bennett, H. P. J., Brubaker, P. L., Seger, M. A., and Solomon, S. (1983). *J. Biol. Chem.* **258,** 8108–8112.
Blake, J., and Li, C. H. (1972). *Biochemistry* **11,** 3459–3461.
Blake, J., and Li, C. H. (1975). *J. Med. Chem.* **18,** 423–426.
Bloomfield, G. A., Scott, A. P., Lowry, P.J., Gilkes, J. J. H., and Rees, L. H. (1974). *Nature (London)* **252,** 492–493.
Bodanszky, M., Klausner, Y., and Ondetti, M. A. (1976). "Peptide Synthesis." Wiley, New York.
Bohus, B., and de Wied, D. (1966). *Science* **153,** 318–320.
Boileau, G., Barbeau, C., Jeannett, L., Chrétien, M., and Drouin, J. (1984). *Nucleic Acids Res.* **11,** 8063–8071.
Boissonnas, R. A., Guttman, St., and Pless, J. (1966). *Experientia* **22,** 526.
Browne, C. A., Bennett, H. P. J., and Solomon, S. (1981). *Biochemistry* **20,** 4538–4546.
Buckley, D. I., Yamashiro, D., and Ramachandran, J. (1981a). *Endocrinology* **109,** 5–9.
Buckley, D. I., Hagman, J., and Ramachandran, J. (1981b). *Endocrinology* **109,** 10–16.
Canova-Davis, E., and Ramachandran, J. (1976). *Biochemistry* **15,** 921–927.
Chang, W.-C., Chung, D., and Li, C. H. (1980a). *Int. J. Pept. Protein Res.* **15,** 261–270.
Chang, A. C. Y., Cochet, M., and Cohen, S. N. (1980b). *Proc. Natl. Acad. Sci. U.S.A.* **77,** 4890–4894.
Chrétien, M., Benjannet, S., Gossard, F., Gianoulakis, C., Crine, P., Lis, M., and Seidah, N. G. (1979). *Can. J. Biochem.* **57,** 1111–1121.
Chung, D., and Li, C. H. (1967). *J. Am. Chem. Soc.* **89,** 4208–4213.
Civelli, O., Birnberg, N., and Herbert, E. (1982). *J. Biol. Chem.* **257,** 6783–6787.
Civelli, O., Douglass, J., and Herbert, E. (1984). In "The Peptides, Analysis, Synthesis, Biology" (S. Udenfriend and J. Meienhofer, eds.), Vol. 6, pp. 69–94. Academic Press, New York.
Cohen, S. N., Chang, A. C. Y., Boyer, H. W., and Helling, R. B. (1973). *Proc. Natl. Acad. Sci. U.S.A.* **70,** 3240–3244.
Cseh, G., Barát, E., Patthy, A., and Gráf, L. (1972). *FEBS Lett.* **21,** 344–346.
Dedman, M. L., Farmer, T. H., and Morris, C. J. O. R. (1955). *Biochem. J.* **59,** xii.
Dedman, M. L., Farmer, T. H., and Morris, C. J. O. R. (1961). *Biochem. J.* **78,** 348–352.
Desaulles, P. A., Riniker, B., and Rittel, W. (1969). *Proc. Int. Symp., Liège 1968* pp. 489–491.
de Wied, D. (1969). In "Frontiers in Neuroendocrinology" (W. F. Ganong and L. Martini, eds.), pp. 97–140. Oxford Univ. Press, London.
de Wied, D. (1977). *Ann. N.Y. Acad. Sci.* **297,** 263–274.
Dixon, H. B. F. (1956). *Biochem. J.* **62,** 25P–26P.
Dixon, H. B. F. (1962). *Biochem. J.* **83,** 91–94.
Dixon, H. B. F., and Weitkamp, L. R. (1962). *Biochem. J.* **84,** 462–468.
Doepfner, W. (1969). *Proc. Int. Congr. Endocrinol., 3rd, Mexico, D. F. 1968* pp. 407–414.
Drouin, J., and Goodman, H. M. (1980). *Nature (London)* **288,** 610–613.
Eberle, A., and Schwyzer, R. (1975). *Helv. Chim. Acta* **58,** 1528–1535.
Eisinger, J. (1969). *Biochemistry* **8,** 3902–3908.
Fauchère, J.-L., and Petermann, C. (1978). *Helv. Chim. Acta* **61,** 1186–1192.
Finn, F. M., Widnell, C. C., and Hofmann, K. (1972). *J. Biol. Chem.* **247,** 5695–5703.
Fujino, M. (1971). *J. Takeda Res. Lab.* **30,** 358–370.

Fujino, M., Hatanaka, C., and Nishimura, O. (1970a). *Chem. Pharm. Bull.* **18,** 771–778.
Fujino, M., Hatanaka, C., Nishimura, O., and Shinagawa, S. (1970b). *Chem. Pharm. Bull.* **18,** 1288–1291.
Fujino, M., Hatanaka, C., and Nishimura, O. (1971a). *Chem. Pharm. Bull.* **19,** 1066–1068.
Fujino, M., Hatanaka, C., Nishimura, O., and Shinagawa, S. (1971b). *Chem. Pharm. Bull.* **19,** 1075–1077.
Geiger, R., and Schröder, H.-G. (1973). *Hoppe-Seyler's Z. Physiol. Chem.* **354,** 156–162.
Geiger, R., Sturm, K., and Siedel, W. (1964a). *Chem. Ber.* **97,** 1207–1213.
Geiger, R., Sturm, K., Vogel, G., and Siedel, W. (1964b). *Z. Naturforsch.* **19B,** 858–860.
Geiger, R., Schröder, H.-G., and Siedel, W. (1969). *Liebigs Ann. Chem.* **726,** 177–187.
Gráf, L., Bajusz, S., Patthy, A., Barát, E., and Cseh, G. (1971). *Acta Biochim. Biophys. Acad. Sci. Hung.* **6,** 415.
Greff, D., Toma, F., Fermandjian, S., Löw, M., and Kisfaludy, L. (1976). *Biochim. Biophys. Acta* **439,** 219–231.
Guillemin, R., Clayton, G. W., Smith, J. D., and Lipscomb, H. S. (1958). *Endocrinology* **63,** 349–358.
Guttmann, St., and Boissonnas, R. A. (1959). *Helv. Chim. Acta* **42,** 1257–1264.
Harris, J. I. (1959). *Nature (London)* **184,** 167–169.
Harris, J. I., and Lerner, A. B. (1957). *Nature (London)* **179,** 1346–1347.
Hoellt, V., Przewlocki, R., and Herz, A. (1978). *Life Sci.* **23,** 1057–1065.
Hofmann, K. (1962). *Ann. Rev. Biochem.* **31,** 213–246.
Hofmann, K., and Yajima, H. (1961). *J. Am. Chem. Soc.* **83,** 2289–2293.
Hofmann, K., and Yajima, H. (1962). *Recent Prog. Horm. Res.* **18,** 41–83.
Hofmann, K., Woolner, M. E., Spühler, G., and Schwartz, E. T. (1958). *J. Am. Chem. Soc.* **80,** 1486–1489.
Hofmann, K., Thompson, T. A., Woolner, M. E., Spühler, G., Yajima, H., Cipera, J. D., and Schwartz, E. T. (1960). *J. Am. Chem. Soc.* **82,** 3721–3726.
Hofmann, K., Yajima, H., Yanaihara, N., Liu, T.-Y., and Lande, S. (1961). *J. Am. Chem. Soc.* **83,** 487–489.
Hofmann, K., Yanaihara, N., Lande, S., and Yajima, H. (1962a). *J. Am. Chem. Soc.* **84,** 4470–4474.
Hofmann, K., Yajima, H., Liu, T.-Y., and Yanaihara, N. (1962b). *J. Am. Chem. Soc.* **84,** 4475–4480.
Hofmann, K., Wells, R. D., Yajima, H., and Rosenthaler, J. (1963). *J. Am. Chem. Soc.* **85,** 1546–1547.
Hofmann, K., Rosenthaler, J., Wells, R. D., and Yajima, H. (1964). *J. Am. Chem. Soc.* **86,** 4991–4999.
Hofmann, K., Andreatta, R., Bohn, H., and Moroder, L. (1970a). *J. Med. Chem.* **13,** 339–345.
Hofmann, K., Wingender, W., and Finn, F. M. (1970b). *Proc. Natl. Acad. Sci. U.S.A.* **67,** 829–836.
Hofmann, K., Montibeller, J. A., and Finn, F. M. (1974). *Proc. Natl. Acad. Sci. U.S.A.* **71,** 80–83.
Holladay, L. A., and Puett, D. (1976). *Biopolymers* **15,** 43–59.
Howard, K. S., Shepherd, R. G., Eigner, E. A., Davies, D. S., and Bell, P. H. (1955). *J. Am. Chem. Soc.* **77,** 3419–3420.
Hruby, V. J., Wilkes, B. C., Cody, W. L., Sawyer, T. K., and Hadley, M. E. (1984). In "Peptide and Protein Reviews" (M. T. W. Hearn, ed.), Vol. 3, pp. 1–64. Dekker, New York and Basel.
Imura, H., Matsuyama, H., Matsukura, S., Miyake, T., and Fukase, M. (1967). *Endocrinology* **80,** 599–602.

Inouye, K., Tanaka, A., and Otsuka, H. (1970a). *Bull. Chem. Soc. Jpn.* **43**, 1163–1172.
Inouye, K., Watanabe, K., Namba, K., and Otsuka, H. (1970b). *Bull. Chem. Soc. Jpn.* **43**, 3873–3882.
Inouye, K., Watanabe, K., Namba, K., Tanaka, A., and Otsuka, H. (1971). *Proc. Symp. Pept. Chem., 8th, Osaka 1970* pp. 95–100.
Inouye, K., Shin, M., and Watanabe, K. (1976a). Jpn. Patent 51-16666 (unexamined).
Inouye, K., Shinozaki, F., Kanayama, M., and Otsuka, H. (1976b). *Bull. Chem. Soc. Jpn.* **49**, 3615–3619.
Inouye, K., Sumitomo, Y., and Shin, M. (1976c). *Bull. Chem. Soc. Jpn.* **49**, 3620–3628.
Inouye, K., Watanabe, K., and Otsuka, H. (1977). *Bull. Chem. Soc. Jpn.* **50**, 211–219.
Inouye, K., Shin, M., Nakamura, M., and Tanaka, A. (1978). *Proc. Symp. Pept. Chem., 15th, Osaka 1977* pp. 177–182.
Jöhl, A., Riniker, B., and Schenkel-Hulliger, L. (1974). *FEBS Lett.* **45**, 172–174.
Kappeler, H., and Schwyzer, R. (1961). *Helv. Chim. Acta* **44**, 1136–1141.
Kappeler, H., Riniker, B., Rittel, W., Desaulles, P., Maier, R., Schär, B., and Staehelin, M. (1967). *Proc. Eur. Pept. Symp., 8th, Noordwijk 1966* pp. 214–220.
Kawauchi, H., and Muramoto, K. (1979). *Int. J. Pept. Protein Res.* **14**, 373–374.
Kawauchi, H., Muramoto, K., and Ramachandran, J. (1978). *Int. J. Pept. Protein Res.* **12**, 318–324.
Kawauchi, A., Abe, K., and Takahashi, A. (1980). *Bull. Jpn. Soc. Sci. Fish.* **46**, 743–747.
Kawauchi, H., Adachi, Y., and Tsubokawa, M. (1981). *Proc. Symp. Pept. Chem., 18th, Nishinomiya 1980* pp. 151–156.
Kawauchi, H., Takahashi, A., and Abe, K. (1982). *Arch. Biochem. Biophys.* **213**, 680–688.
Kisfaludy, L., Löw, M., Szirtes, T., Schön, I., Sárközi, M., Bajusz, S., Turan, A., Juhász, A., Beke, R., Gráf, L., and Medzihradszky, K. (1972). *Proc. Am. Pept. Symp., 3rd, Boston* pp. 299–303.
Kloppenborg, P. W. C., Island, D. P., Liddle, G. W., Michelakis, A. M., and Nicholson, W. E. (1968). *Endocrinology* **82**, 1053–1058.
Koyama, K., Kawatani, H., Yajima, H., Fujino, M., and Nishimura, O. (1976a). *Chem. Pharm. Bull.* **24**, 2106–2111.
Koyama, K., Watanabe, H., Kawatani, H., Iwai, J., and Yajima, H. (1976b). *Chem. Pharm. Bull.* **24**, 2558–2563.
Krieger, D. T., Liotta, A. S., Brownstein, M. J., and Zimmerman, E. A. (1980). *Recent Prog. Horm. Res.* **36**, 277–344.
Kumar, S. (1975). *Biochem. Biophys. Res. Commun.* **66**, 1063–1068.
Larsson, L.-I. (1977). *Lancet* **2**, 1321–1323.
Larsson, L.-I. (1978). *Histochemistry* **55**, 225–233.
Lebovitz, H. E., and Engel, F. L. (1964). *Metabol. Clin. Exp.* **13**, 1230–1246.
Lee, T. H., and Lerner, A. B. (1956). *J. Biol. Chem.* **221**, 943–959.
Lee, T. H., Lerner, A. B., and Buettner-Janusch, V. (1959). *J. Am. Chem. Soc.* **81**, 6084.
Lee, T. H., Lerner, A. B., and Buettner-Janusch, V. (1961). *J. Biol. Chem.* **236**, 2970–2974.
Li, C. H. (1962). *Recent Prog. Horm. Res.* **18**, 1–40.
Li, C. H. (1964). *Nature (London)* **201**, 924–925.
Li, C. H. (1972). *Biochem. Biophys. Res. Commun.* **49**, 835–839.
Li, C. H., and Chung, D. (1976). *Proc. Natl. Acad. Sci. U.S.A.* **73**, 1145–1148.
Li, C. H., and Dixon, J. S. (1956). *Science* **124**, 934.
Li, C. H., Geschwind, I. I., Levy, A. L., Harris, J. I., Dixon, J. S., Pon, N. G., and Porath, J. O. (1954). *Nature (London)* **173**, 251–253.
Li, C. H., Geschwind, I. I., Cole, R. D., Raake, I. D., Harris, J. I., and Dixon, J. S. (1955a). *Nature (London)* **176**, 687–689.

1. ACTH: STRUCTURE–FUNCTION RELATIONSHIP

Li, C. H., Geschwind, I. I., Dixon, J. S., Levy, A. L., and Harris, J. I. (1955b). *J. Biol. Chem,* **213,** 171–185.
Li, C. H., Dixon, J. S., and Chung, D. (1958). *J. Am. Chem. Soc.* **80,** 2587–2588.
Li, C. H., Meienhofer, J., Schnabel, E., Chung, D., Lo, T.-B., and Ramachandran, J. (1960). *J. Am. Chem. Soc.* **82,** 5760–5762.
Li, C. H., Meienhofer, J., Schnabel, E., Chung, D., Lo, T.-B., and Ramachandran, J. (1961). *J. Am. Chem. Soc.* **83,** 4449–4457.
Li, C. H., Chung, D., Ramachandran, J., and Gorup, B. (1962). *J. Am. Chem. Soc.* **84,** 2460–2462.
Li, C. H., Barnafi, L., Chrétien, M., and Chung, D. (1965). *Nature (London)* **208,** 1093–1094.
Li, C. H., Chung, D., Oelofsen, W., and Naudé, R. J. (1978). *Biochem. Biophys. Res. Commun.* **81,** 900–906.
Liotta, A., Osathanondh, R., Ryan, K. J., and Krieger, D. T. (1977). *Endocrinology* **101,** 1552–1558.
Lipscomb, H. S., and Nelson, D. M. (1962). *Endocrinology* **71,** 13–23.
Löw, M., Kisfaludy, L., and Fermandjian, S. (1975). *Acta Biochim. Biophys. Acad. Sci. Hung.* **10,** 229–231.
Löw, M., Kisfaludy, L., Hajos, G., Szporny, L., Mihály, K., Makara, G. B., Toma, F., Dive, V., and Fermandjian, S. (1980). *Proc. Eur. Pept. Symp., 16th, Helsingør* pp. 513–519.
Lowry, P. J., Bennett, H. P. J., McMartin, C., and Scott, A. P. (1974). *Biochem. J.* **141,** 427–437.
Mains, R. E., Eipper, B. A., and Ling, N. (1977). *Proc. Natl. Acad. Sci. U.S.A.* **74,** 3014–3018.
Maxam, A. M., and Gilbert, W. (1977). *Proc. Natl. Acad. Sci. U.S.A.* **74,** 560–564.
Moyle, W. R., Kong, Y.-C., and Ramachandran, J. (1973). *J. Biol. Chem.* **248,** 2409–2417.
Nakamura, M. (1972). *J. Biochem.* **71,** 1029–1041.
Nakanishi, S., Taii, S., Hirata, Y., Matsukura, S., Imura, H., and Numa, S. (1976). *Proc. Natl. Acad. Sci. U.S.A.* **73,** 4319–4323.
Nakanishi, S., Inoue, A., Taii, S., and Numa, S. (1977). *FEBS Lett.* **84,** 105–109.
Nakanishi, S., Inoue, A., Kita, T., Nakamura, M., Chang, A. C. Y., Cohen, S. N., and Numa, S. (1979). *Nature (London)* **278,** 423–427.
Ney, R. L., Ogata, E., Shimizu, N., Nicholson, W. E., and Liddle, G. W. (1964). *Proc. Int. Congr. Endocrinol., 2nd, London* pp. 1184–1191.
Ng, T. B., Chung, D., and Li, C. H. (1981). *Int. J. Pept. Protein Res.* **18,** 443–450.
Nishimura, O., Hatanaka, C., and Fujino, M. (1975). *Chem. Pharm. Bull.* **23,** 1212–1220.
Notake, M., Tobimatsu, T., Watanabe, Y., Takahashi, H., Mishina, M., and Numa, S. (1983). *FEBS Lett.* **156,** 67–71.
Okamoto, K., Yasumura, K. Shimamura, S., Nakamura, M., Tanaka, A., and Yajima, H. (1979). *Chem. Pharm. Bull.* **27,** 499–507.
Orwoll, E. S., and Kendall, J. W. (1980). *Endocrinology* **107,** 438–442.
Otsuka, H., and Inouye, K. (1964). *Bull. Chem. Soc. Jpn.* **37,** 1465–1471.
Otsuka, H., and Inouye, K. (1975). *Pharmacol. Ther. B* **1,** 501–527.
Otsuka, H., Inouye, K., Kanayama, M., and Shinozaki, F. (1965a). *Bull. Chem. Soc. Jpn.* **38,** 679–680.
Otsuka, H., Inouye, K., Kanayama, M., and Shinozaki, F. (1965b). *Bull. Chem. Soc. Jpn.* **38,** 1563–1564.
Otsuka, H., Inouye, K., Shinozaki, F., and Kanayama, M. (1965c). *J. Biochem.* **58,** 512–514.

Otsuka, H., Shin, M., Kinomura, Y., and Inouye, K. (1970). *Bull. Chem. Soc. Jpn.* **43**, 196–200.
Pickering, B. T., and Li, C. H. (1963). *Biochim. Biophys. Acta* **74**, 156–157.
Ramachandran, J. (1970). *Biochem. Biophys. Res. Commun* **41**, 353–357.
Ramachandran, J. (1973). *In* "Hormonal Proteins and Peptides" (C. H. Li, ed.), Vol. 2, pp. 1–28. Academic Press, New York.
Ramachandran, J., and Li, C. H. (1967). *Adv. Enzymol.* **29**, 391–477.
Ramachandran, J., Chung, D., and Li, C. H. (1965). *J. Am. Chem. Soc.* **87**, 2696–2708.
Riniker, B. (1971). *Proc. Int. Symp., 2nd, Liège* pp. 519–520.
Riniker, B., and Rittel, W. (1970). *Helv. Chim. Acta* **53**, 513–519.
Riniker, B., Sieber, P., Rittel, W., and Zuber, H. (1972). *Nature (London) New Biol.* **235**, 114–115.
Roberts, J. L., and Herbert, E. (1977a). *Proc. Natl. Acad. Sci. U.S.A.* **74**, 4826–4830.
Roberts, J. L., and Herbert, E. (1977b). *Proc. Natl. Acad. Sci. U.S.A.* **74**, 5300–5304.
Rubinstein, M., Stein, S., and Udenfriend, S. (1978). *Proc. Natl. Acad. Sci. U.S.A.* **75**, 669–671.
Saffran, M., and Schally, A. V. (1955). *Endocrinology* **56**, 523–532.
Saito, E., and Odell, W. D. (1983). *Proc. Natl. Acad. Sci. U.S.A.* **80**, 3792–3796.
Sanger, F., and Coulson, A. R. (1978). *FEBS Lett.* **87**, 107–110.
Sawyer, T. K., Hruby, V. J., Darman, P. S., and Hadley, M. E. (1982). *Proc. Natl. Acad. Sci. U.S.A.* **79**, 1751–1755.
Sayers, M. A., Sayers, G., and Woodbury, L. A. (1948). *Endocrinology* **42**, 379–393.
Sayers, G., Swallow, R. L., and Giordano, N. D. (1971). *Endocrinology* **88**, 1063–1068.
Sayers, G., Seelig, S., Kumar, S., Karlaganis, G., Schwyzer, R., and Fujino, M. (1974). *Proc. Soc. Exp. Biol. Med.* **145**, 176–181.
Schröder, E., and Lübke, K. (1966). *In* "The Peptides," Vol. 2, pp. 194–251. Academic Press, New York.
Schwyzer, R. (1964). *Annu. Rev. Biochem.* **33**, 259–286.
Schwyzer, R. (1977). *Ann. N.Y. Acad Sci.* **297**, 3–26.
Schwyzer, R., and Kappeler, H. (1963). *Hlv. Chim. Acta* **46**, 1550–1572.
Schwyzer, R., and Li, C. H. (1958). *Nature (London)* **182**, 1669–1670.
Schwyzer, R., and Sieber, P. (1963). *Nature (London)* **199**, 172–174.
Schwyzer, R., and Sieber, P. (1966). *Helv. Chim. Acta* **49**, 134–158.
Schwyzer, R., Costopanadiotis, A., and Sieber, P. (1963). *Helv. Chim. Acta* **44**, 870–889.
Schwyzer, R., Schiller, P., Seelig, S., and Sayers, G. (1971). *FEBS Lett.* **19**, 229–231.
Scott, A. P., and Lowry, P. J. (1974). *Biochem. J.* **139**, 593–602.
Scott, A. P., Ratcliffe, J. G., Rees, L. H., Landon, J., Bennett, H. P. J., Lowry, P. J., and McMartin, C. (1973). *Nature (London) New Biol.* **244**, 65–67.
Scott, A. P., Lowry, P. J., Ratcliffe, J. G., Rees, L. H., and Landon, J. (1974). *J. Endocrinol.* **61**, 355–367.
Seelig, S., and Sayers, G. (1973). *Arch. Biochem. Biophys.* **154**, 230–239.
Seelig, S., Sayers, G., Schwyzer, R., and Schiller, P. (1971). *FEBS Lett.* **19**, 232–234.
Seelig, S., Kumar, S., and Sayers, G. (1972). *Proc. Soc. Exp. Biol. Med.* **139**, 1217–1219.
Shepherd, R. G., Howard, K. S., Bell, P. H., Cacciola, A. R., Child, R. G., Davies, M. C., English, J. P., Finn, B. M., Meisenhelder, J. H., Moyer, A. W., and van der Scheer, J. (1956). *J. Am. Chem. Soc.* **78**, 5051–5059.
Shimamura, S., Okamoto, K., Nakamura, M., Tanaka, A., and Yajima, H. (1978). *Int. J. Pept. Protein Res.* **12**, 170–172.
Shin, M., Inouye, K., and Otsuka, H. (1971). *Proc. Symp. Pept. Chem., 8th, Osaka 1970* pp. 91–94.

Shizume, K., Lerner, A. B., and Fitzpatrick, T. B. (1954). *Endocrinology* **54**, 553–560.
Sieber, P., Rittel, W., and Riniker, B. (1972). *Helv. Chim. Acta* **55**, 1243–1248.
Soma, G., Kitahara, N., Nishizawa, T., Nanami, H., Kotake, C., Okazaki, H., and Andoh, T. (1984). *Nucleic Acids Res.* **12**, 8029–8041.
Szporny, L., Hajós, Gy. T., Szeberényi, Sz., and Fekete, Gy. (1969). *Proc. Int. Symp., Liège 1968* pp. 492–494.
Tanaka, A. (1971). *Endocrinol. Jpn.* **18**, 155–168.
Tesser, G. I., and Buis, J. T. (1971). *Rec. Trav. Chim.* **90**, 444–457.
Tesser, G. I., and Rittel, W. (1969). *Rec. Trav. Chim.* **88**, 553–561.
Tesser, G. I., and Schwyzer, R. (1966). *Helv. Chim. Acta* **49**, 1013–1022.
Tesser, G. I., Maier, R., Schenkel-Hulliger, L., Barthe, P. L., Kamber, B., and Rittel, W. (1973a). *Acta Endocrinol.* **74**, 56–66.
Tesser, G. I., Pleumekers, A. W. J., Bassie, W., and Balvert-Geers, I. C. (1973b). *Rec. Trav. Chim.* **92**, 1210–1222.
Uhler, M., and Herbert, E. (1983). *J. Biol. Chem.* **258**, 257–261.
Van Nispen, J. W., and Greven, H. M. (1982). *Pharmacol. Ther.* **16**, 67–102.
Van Nispen, J. W., and Tesser, G. I. (1975). *Int. J. Pept. Protein Res.* **7**, 57–67.
Van Nispen, J. W., and Greven, H. M. (1982). *Pharmacol. Ther.* **16**, 67–102.
(1977a). *Acta Endocrinol.* **84**, 470–484.
Van Nispen, J. W., Tesser, G. I., and Nivard, R. J. F. (1977b). *Int. J. Pept. Protein Res.* **9**, 193–202.
Vernikos-Danellis, J., Anderson, E., and Trigg, L. (1966). *Endocrinology* **79**, 624–630.
Waller, J. P., and Dixon, H. B. F. (1960). *Biochem. J.* **75**, 320–328.
White, W. F. (1953). *J. Am. Chem. Soc.* **75**, 503–504.
White, W. F. (1955). *J. Am. Chem. Soc.* **77**, 4691–4692.
Yajima, H., Koyama, K., Kiso, Y., Tanaka, A., and Nakamura, M. (1976). *Chem. Pharm. Bull.* **24**, 492–499.
Yamashiro, D., and Li, C. H. (1973). *J. Am. Chem. Soc.* **95**, 1310–1315.
Yasumura, K., Okamoto, K., Shimamura, S., Nakamura, M., Odaguchi, K., Tanaka, A., and Yajima, H. (1982). *Chem. Pharm. Bull.* **30**, 866–873.
Yoshimoto, Y., Wolfsen, A. R., Hirose, F., and Odell, W. D. (1977). *Science* **197**, 575–577.

ACTH Receptors

J. RAMACHANDRAN
Genentech, Inc.
South San Francisco, California 94080, and
Department of Biochemistry and Biophysics
University of California
San Francisco, California 94143

I. Introduction

Acute stimulation of glucocorticoid synthesis in the zona fasciculata of the adrenal cortex and the long-term maintenance of the steroidogenic capacity of the adrenal cortex are the primary functions of the pituitary hormone corticotropin (ACTH). In addition to these, ACTH exerts several extraadrenal actions. The darkening of the amphibian melanophores through melanin dispersion and stimulation of lipolysis in adipocytes are the most notable of the many effects induced by ACTH in the absence of adrenals. ACTH exerts all its actions by interacting with specific receptors on the external surface of the target cell. This initial interaction results in the activation of the enzyme adenylate cyclase located on the cytoplasmic side of the adrenocortical cell, adipocyte, or melanophore plasma membrane. The consequent increase in intracellular cAMP is generally credited with mediating the cell-specific physiological responses to the hormone.

Although the second messenger concept and the involvement of cAMP were formulated partly on the basis of studies of the action of ACTH on the adrenal cortex (Haynes and Berthet, 1957), there has been considerable debate as to the nature of the second messenger mediating the steroidogenic action of ACTH. Doubts about the role of cAMP in ACTH-induced steroidogenesis first surfaced in 1972 when it was observed that physiological concentrations of the hormone could stimulate corticosterone formation in isolated rat adrenocortical cells without causing detectable changes in the intracellular concentration of cAMP (Beall and Sayers, 1972; Mackie *et al.*, 1972; Nakamura *et al.*, 1972; Moyle *et al.*, 1973). Studies with analogs of ACTH also emphasized the discrepancy between the peptide concentrations required for eliciting steroid synthesis vs those needed for stimulating cAMP formation (Ramachandran *et al.*, 1976; Ramachandran and Moyle, 1977). Furthermore, both cGMP (Harrington *et al.*, 1978; Perchellet *et al.*, 1978) and calcium (Neher and Milani, 1978; Perchellet and Sharma, 1979; Rubin and Laychock, 1978; Yanagibashi *et al.*, 1978) have been proposed as second messengers involved in the stimulation of steroidogenesis by ACTH. The first binding studies using ^{125}I-labeled ACTH preparations (Lefkowitz *et al.*, 1971) were interpreted as evidence for the presence of two classes of binding sites for ACTH in adrenocortical cells. This report also served to fuel the debate about the precise role of cAMP in the actions of ACTH. The difficulties encountered in detecting and characterizing the physiologically relevant ACTH receptors contributed, to a large extent, to the controversies concerning the mechanisms by which ACTH regulates adrenocortical function.

In this chapter the early binding studies and the problems associated with studies of ACTH receptors are discussed. The recent preparation of a well-characterized ^{125}I-derivative of ACTH with full biological potency and the unambiguous identification and characterization of ACTH receptors in rat adrenocortical cells is then discussed. The role of extracellular calcium and the characterization of ACTH receptors in fetal and adult human adrenocortical cells are reviewed. ACTH receptors in rat adipocytes and 3T3-L1 cells are also discussed.

II. Detection of ACTH Receptors

A. HISTORICAL SURVEY

The presence of ACTH receptors on the plasma membrane of adrenocortical cells was inferred from the finding that brief exposure of adrenal tissue to ACTH leads to a persistent increase in steroid production, and that the effect is blocked by treatment with trypsin or anti-ACTH antibodies after brief incubation with the hormone (Taunton *et al.*, 1967). Several investigators also presented evidence that ACTH does not have to enter the adrenocortical cell in order to stimulate steroid synthesis. These early studies have been thoroughly reviewed by Halkerston (1975).

The first binding studies of ACTH receptors using a radioactive ligand were reported by Lefkowitz *et al.* (1970b). Using an ^{125}I-labeled preparation of porcine ACTH, these authors reported that the radioligand bound specifically to extracts of a mouse adrenal tumor. On the basis of Scatchard analysis of the interaction of the ^{125}I-labeled ACTH preparation with the mouse adrenal tumor extract, Lefkowitz et al. (1971) suggested that there are two classes of ACTH binding sites on the adrenocortical cell: 60 high-affinity sites with an apparent dissociation constant (K_d) of 1.1×10^{-12} M and 360,000 low-affinity sites with a K_d of 3.3×10^{-8} M. Lefkowitz et al. (1970a) also reported that the interaction of ACTH with adrenal receptors does not require calcium. Hofmann et al. (1970) investigated the binding of ([^{14}C]Phe^7, Gln^5)-ACTH (1–20) amide to a particulate preparation from a bovine adrenal cortex and reported correlation between binding and corticotropic activity measured in vivo. This group could detect only low-affinity binding sites (Finn et al., 1972). Several laboratories have employed ^{125}I-labeled ACTH preparation to investigate binding sites in crude membrane preparations derived from rat (Saez et al., 1974; Ways et al., 1976), ovine (Saez et al., 1974), human (Saez et al., 1974; Nishisato, 1980), and rabbit (Durand and Locatelli, 1980) adrenal glands. In all cases only low-affinity binding sites of high capacity were detected.

Binding of ^{125}I-labeled ACTH to intact, isolated rat adrenocortical cells was first reported by McIlhinney and Schulster in 1975. These authors also found two classes of binding sites, a high-affinity site with $K_d = 2.5 \times 10^{-10}$ M (3,000 sites/cell) and a low-affinity site with $K_d = 1 \times 10^{-8}$ M (30,000 sites/cell). McIlhinney and Schulster (1975) suggested that the high-capacity, low-affinity binding sites is coupled to the adenylate cyclase system. Similar results were reported by Yanagibashi et al. (1978) who estimated that the rat adrenocortical cells contain 7350 high-affinity sites ($K_d = 2.6 \times 10^{-10}$ M) and 57,400 low affinity sites ($K_d = 7.1 \times 10^{-9}$ M). Yanagibashi et al. (1978) also reported that the apparent dissociation constants derived from the effects of ACTH on steroidogenesis and calcium influx correlated well with the K_d of the high-affinity receptor. These authors concluded that the high-affinity binding site is linked to calcium influx and consequent regulation of steroidogenesis at physiological concentrations of ACTH, whereas the low-affinity site is coupled to adenylate cyclase which is activated by supraphysiological concentrations of the hormone. As mentioned earlier, the concept of two receptor populations in adrenocortical cells was also suggested by numerous studies which showed an apparent dissociation between steroidogenesis and cAMP production induced by ACTH (Beall and Sayers, 1972; Nakamura et al., 1972; Mackie et al., 1972) as well as by numerous analogs of the hormone (Moyle et al., 1973; Ramachandran et al., 1976; Ramachandran and Moyle, 1977).

B. BIOLOGICAL ACTIVITY OF ^{125}I-LABELED ACTH

The major problems encountered in direct binding studies of ACTH receptors can be traced to the generally new biological potencies of the ^{125}I-labeled ACTH preparations employed in the investigations noted above. Although most investigations reported that the radioligands employed in binding studies were biologically active, these radioactive ACTH preparations generally had much less than 50% biological potency which could often be ascribed to the unmodified "cold" hormone present in the preparation. Lefkowitz et al. (1970b) separated the ^{125}I-labeled ACTH from uniodinated ACTH by chromatography on carboxymethylcellulose and reported a biological potency of 32 units/mg in the adenylate cyclase assay. Although the ACTH preparation used by Lefkowitz et al. for iodination had an activity of only 64 units/mg, purified ACTH has been reported to exhibit a potency of 200 units/mg in the adenylate cyclase assay (Rae and Schimmer, 1974). Thus, even the purified ^{125}I-labeled ACTH employed by Lefkowitz et al. (1970b) had only 16% of the potency of ACTH. Ways et al. (1976) report that the ^{125}I-labeled ACTH employed in their studies contained 0.2 atoms of iodine per mole of peptide, and this preparation had a biological activity of 15 units/mg. Since only 20% of the molecules could carry an iodine atom in this preparation, the low biological activity may well have been due to the 80% unmodified ACTH. Even if all the iodinated molecules were inactive, the preparation would be expected to possess 80% biological activity. That the activity is as low as 7.5% suggests that, beside the introduction of the bulky iodine atom, the conditions used in the iodination procedure were causing inactivation of the hormone. Previous reports did, indeed, suggest that iodination of ACTH with chloramine-T resulted in major loss of biological activity (Greenwood et al., 1963; Landon et al., 1967; Rees et al., 1971). Rae and Schimmer (1974) showed that monoiodo derivatives of ACTH possessed only 30% of the potency of the hormone; however, the activity could be restored by reduction with cysteine, suggesting that the oxidation of the single methionine residue in ACTH under the iodinating conditions was responsible for the decrease in biological activity. Dedman et al. (1961) had shown previously that oxidation of the methionine residue in ACTH lowers the biological potency to less than 10% of the native hormone. Rae and Schimmer (1974) also found that iodination with lactoperoxidase instead of chloramine-T did not preserve hormonal activity.

Structure–activity relationship studies have shown that introduction of the bulky iodine atom into the tyrosine residue in position 2 also causes a drastic reduction in biological potency. Lowry et al. (1973) and Lemaire et al. (1977) prepared ACTH (1–24) and ACTH (1–39), respectively, in which the tyrosine residues in positions 2 and/or 23 were replaced by 3,5-diiodotyrosine residues. Replacement of Tyr^2 by the diiodo derivative resulted in a 98% decrease in biological potency, whereas replacement of Tyr^{23} had a much less detrimental effect. Both the chloramine-T method (Lefkowitz et al., 1970b) and the lactoperoxidase procedure (McIlhinney and Schulster, 1974) preferentially iodinate Tyr^2. Thus, the standard iodination conditions employed for preparing ^{125}I-labeled ACTH used in all the binding studies cited above led to radioligands of low biological potency.

C. Studies with [^{3}H]ACTH

In order to avoid the problems encountered in the preparation of ^{125}I-labeled ACTH, some investigators turned to the use of ^{14}C and ^{3}H as the radioactive tracers. Hofmann et al. (1970) used synthetic ([^{14}C]Phe^7, Gln^5)-ACTH (1–20) amide which had a specific radioactivity of 0.126 mCi/mmol compared to 2000 Ci/mmol for ^{125}I-labeled ACTH. The ^{14}C-labeled peptide had a specific radioactivity seven order of magnitude lower than that of ^{125}I-labeled ACTH and is clearly of little use in binding studies. Much higher specific radioactivities can be obtained with tritium. Schwyzer and Karlaganis (1973) prepared the analog Phe^2,4,5-dehydro[4,5-^{3}H]norvaline^4-ACTH (1–24) with a specific radioactivity of 7.42 Ci/mmol, but this tritiated peptide possessed only 10% of the biological potency of ACTH (1–24). Brundish and Wade (1973) and Ramachandran and Behrens (1977) prepared [3,5-^{3}H]Tyr^{23}-ACTH (1–24) and [3,5-^{3}H]Tyr^{23}-ACTH (1–39), respectively, by catalytic dehalogenation of the corresponding 3,5-diiodotyrosine derivative in the presence of pure tritium. Both tritiated peptides had a specific radioactivity of 46 Ci/mmol and full biological potency. Ramachandran et al. (1980a) also prepared [3,5-^{3}H]Tyr^{2,23}-ACTH (1–39) containing two tritium atoms in each tyrosine residue. This peptide had a specific radioactivity of 90 Ci/mmol and was physicochemically and biologically indistinguishable from synthetic human ACTH.

Ramachandran et al. (1980b) used this [^{3}H]ACTH to investigate ACTH receptors in isolated rat adrenocortical cells. Owing to the relatively lower specific radioactivity of [^{3}H]ACTH compared to that of ^{125}I-labeled ACTH, binding studies were performed at a [^{3}H]ACTH concentration of 3 nM. Under these conditions there was considerable binding of the tritiated hormone to nonreceptor components since significant amounts of radioactivity bound to the cells were displaced by unrelated hormones and basic peptides such as polylysine. Since polylysine preparations did not stimulate steroidogenesis or cAMP formation in the rat adrenocortical cells and had no effect on the ability of ACTH to stimulate steroid synthesis and cAMP formation, binding studies were performed in the presence of 0.01% polylysine. In the presence of polylysine, binding of [^{3}H]ACTH was highly specific (Ramachandran *et al.*, 1980b). These studies suggested the presence of a single class of binding sites with a K_d of 2–4 nM and a capacity of 4000 sites/cell. Maximal steroidogenesis occurred at concentrations of [^{3}H]ACTH at which no binding or cAMP production could be detected. The binding of [^{3}H]ACTH appeared to correlate with cAMP formation which was half-maximal at an ACTH concentration of 2–4 nM. These studies showed that the adrenocortical cells contain a large number of nonfunctional sites in addition to a limited number of high-affinity sites that appear to be related to cAMP formation. Although higher affinity sites which may interact with ACTH at lower, steroidogenic concentrations of the hormone could not be detected with the tritiated hormone, such sites could not be ruled out owing to the intrinsic limitation of the specific radioactivity of the [^{3}H]ACTH.

D. SYNTHESIS OF A RADIOLIGAND WITH FULL BIOLOGICAL POTENCY

The two major causes of inactivation during chloramine-T- or lactoperoxidase-catalyzed iodination of ACTH are the preferential introduction of iodine into the tyrosine residue in position 2 and the oxidation of the side-chain of the methionine residue in position 4. These difficulties encountered during iodination were circumvented by the synthesis of an analog of the hormone in which the tyrosine residue in position 2 is replaced by a phenylalanine residue and the methionine in position 4 is replaced by a norleucine residue (Buckley *et al.*, 1981a). This peptide, Phe^2, Nle^4-ACTH (1–38) was synthesized by the solid-phase peptide synthesis procedure and purified by partition chromatography. Phe^2, Nle^4-ACTH (1–38) was found to be equipotent with ACTH in stimulating steroidogenesis in adrenocortical cells.

By eliminating the tyrosine residue in position 2, iodination can be restricted to the tyrosine residue in position 23, which lies outside the amino acid sequence known to be highly important for biological activity (Ramachandran, 1973). Substitution of norleucine for methionine in position 4 eliminated a potential site of oxidation during iodination. Thus, iodination of Phe^2, Nle^4-ACTH (1–38) resulted in the formation of only two products, a monoiodo and a diiodo derivative (Buckley *et al.*, 1981a). These were readily separated from each other and unmodified Phe^2, Nle^4-ACTH (1–38) by reverse-phase HPLC. The monoiodo derivative of Phe^2,

Nle^4-ACTH (1–38) was found to be as potent as ACTH in stimulating steroidogenesis.

^{125}I-labeled Tyr^{23}, Phe^2, Nle^4-ACTH (1–38) (referred to hereafter as ^{125}I-ACTH analog) was prepared by a modified chloramine-T procedure and isolated in a homogeneous state by reverse-phase HPLC (Buckley et al., 1981b). ^{125}I-ACTH analog was completely separated from unmodified Phe^2,Nle^4-ACTH (1–38) (Fig. 1). This homogeneous radioligand stimulated corticosterone production in isolated rat adrenocortical cells to the same maximal rate as ACTH and was found to be equipotent with the hormone (Fig. 2). The specific radioactivity of ^{125}I-ACTH analog was determined unambiguously by comparing the antiserum binding curve for ^{125}I-ACTH analog with the binding curve for [^{3}H]ACTH to the same dilution of ACTH antiserum. Since the specific radioactivity of the [^{3}H]ACTH preparation was known from direct concentration and radioactivity measurements (Ramachandran and Behrens, 1977), the specific radioactivity of ^{125}I-ACTH analog could be obtained from the concentration required for half-maximal saturation of the antiserum (Buckley et al., 1981b). ^{125}I-ACTH analog was found to have a specific radioactivity of 1800 ±75 Ci/mmol, which is very close to the theoretically expected value for the introduction of a single iodine atom into the peptide, taking into account the purity and age of the radioisotope.

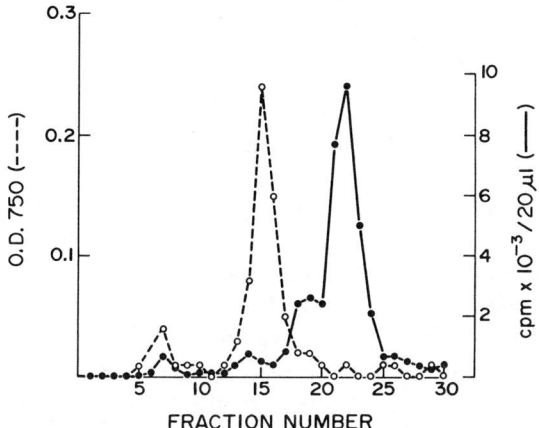

FIG. 1. Reverse-phase HPLC of a mixture of Phe^2, Nle^4-ACTH (1–38) and ^{125}I-labeled Tyr^{23}, Phe^2, Nle^4-ACTH (1–38). A mixture of Phe^2, Nle^4-ACTH (1–38) and ^{125}I-labeled Tyr^{23}, Phe^2, Nle^4-ACTH (1–38) was applied on an Altex Lichrosorb C-8 column and eluted isocratically with pyridine acetate, pH 5.5, containing 14% 1-propanol. Flow rate 0.5 ml/min; fractions, 0.5 ml. Detection was performed using the Lowry procedure on 0.1 ml aliquots and by counting 20 µl aliquots in a gamma counter.

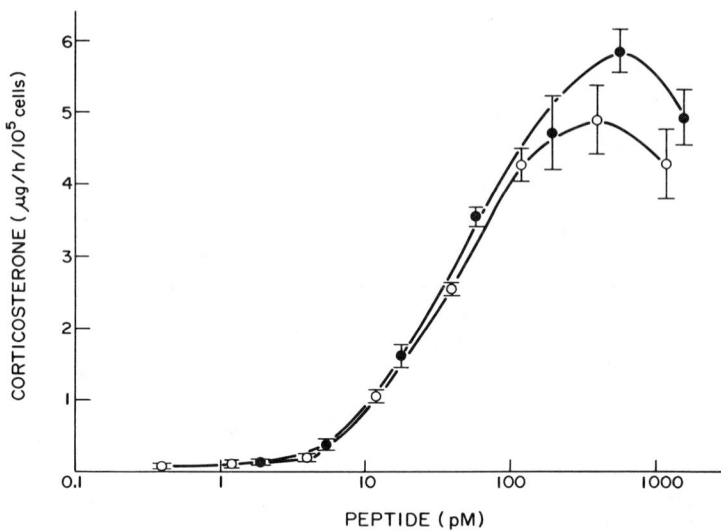

FIG. 2. Biological activity of ^{125}I-ACTH analog. HPLC-purified ^{125}I-ACTH analog (○) or ACTH (●) were incubated with isolated rat adrenocortical cells for 1 hr at 37°C and corticosterone production was monitored by a specific RIA.

III. Characterization of ACTH Receptors in Rat Adrenocortical Cells

A. Binding Characteristics

Buckley and Ramachandran (1981) used the ^{125}I-ACTH analog with full biological activity and high specific radioactivity to investigate ACTH receptors in rat adrenocortical cells. Binding of ^{125}I-ACTH was both ligand and cell specific (Table I). Approximately 85–90% of the radioactivity bound to adrenocortical cells was displaced by 44 nM unlabeled ACTH but not by other basic peptides like β-endorphin, insulin, or polylysine. No specific binding was observed with rat Leydig cells. Under the conditions employed in this study, there was no significant degradation of the radioligand. Binding of ^{125}I-ACTH analog to rat adrenocortical cells was rapid at room temperature and was maximal by 30 min. The dissociation of ^{125}I-ACTH analog from adrenocortical cells followed first order kinetics with a half-time dissociation of 40 min at 23°C.

A linear relationship (correlation coefficient, 0.997) was observed between the number of adrenocortical cells and the amount of ^{125}I-ACTH analog specifically bound (Fig. 3). The affinity and capacity of binding at equilibrium were determined from analyses of the competitive inhibition

2. ACTH RECEPTORS

Table I—Specificity of Binding of ^{125}I-ACTH Analog[a]

Displacing ligand	^{125}I-ACTH analog bound[b] (cpm)
Rat adrenocortical cells	
None	8292 ± 365
ACTH (0.2 µg/ml)	1081 ± 242
β-Endorphin (0.2 µg/ml)	9400 ± 339
Insulin (0.2 µg/ml)	9335 ± 146
Polylysine (0.2 µg/ml)	9017 ± 357
Rat Leydig cells	
None	1057 ± 88
ACTH (0.2 µg/ml)	1072 ± 92
Polylysine (0.2 µg/ml)	1073 ± 74

[a] Cells were incubated with ^{125}I-ACTH analog (193,370 cpm) in 0.5 ml of medium 199/0.5% albumin in the presence or absence of displacing ligands at 23°C. Each incubation tube contained adrenocortical cells representing 16.4 ± 0.7 µg of DNA or Leydig cells representing 120.3 ± 2.43 µg of DNA.

[b] Values are mean ± SEM for duplicate analyses of duplicate incubations.

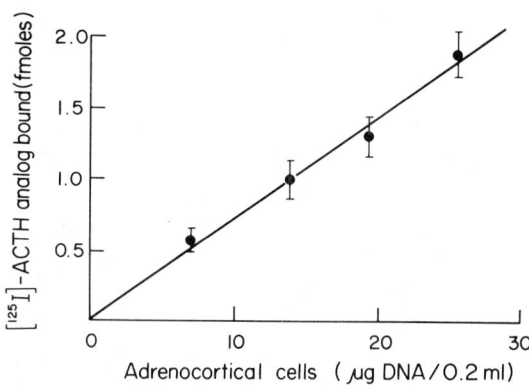

FIG. 3. Relationship of binding of ^{125}I-ACTH analog to cell number. Various amounts of adrenal cells were incubated with ^{125}I-ACTH analog (0.2 nM) at 23°C for 1 hr. Cells were separated by centrifugation through a bovine serum albumin cushion and processed as described by Buckley and Ramachandran (1981). Nonspecific binding to cells measured in the presence of 444 nM ACTH has been subtracted. Values are the mean ± SEM for triplicate incubations.

of binding of ^{125}I-ACTH analog by unlabeled ACTH. This approach is valid because the biological potencies of ACTH and ^{125}I-ACTH analog are identical (Buckley et al., 1981b). In addition, the concentration of nonradioactive iodinated analog required for 50% inhibition of binding of ^{125}I-ACTH analog was 1.26 ± 0.19 nM (mean ± SE) compared to a value of 1.35 ± 0.19 nM for ACTH. Increasing concentrations of ACTH produced a binding inhibition curve which fits a computed line for a single class of binding sites (Fig. 4, top). Scatchard analysis of the data (Fig. 4, bottom) also shows that the data are best fitted by a single straight line. Analysis of the binding data from several experiments yielded a value of 1.41 ± 0.21 nM for the apparent K_d. The number of sites/cell was estimated to be 3840 ± 1045. These results are at variance with the conclusions of previous binding studies performed with ^{125}I-labeled ACTH preparations of low biological potencies (Lefkowitz et al., 1971) but are in excellent agreement with the results of earlier binding studies using [^{3}H]ACTH with a specific radioactivity of 90 Ci/mmol and full biological potency (Ramachandran et al., 1980b). The results obtained using ^{125}I-ACTH analog with full biological potency and the highest specific radioactivity to date suggest that it is reasonable to conclude that there is only one class of specific binding sites for ACTH in adrenocortical cells.

B. CORRELATION OF BINDING WITH cAMP SYNTHESIS AND STEROIDOGENESIS

As mentioned in Section I, the concentrations of ACTH required for half-maximal stimulation of steroidogenesis and cAMP production in adrenocortical cells are vastly different. Physiological concentrations of the hormone which stimulate steroidogenesis maximally are capable of eliciting only marginal increases in cAMP levels. The relationship between the single class of specific binding sites revealed by the studies using ^{125}I-ACTH analog and the physiological responses have been investigated (Buckley and Ramachandran, 1981). As seen from the results shown in Fig. 5, specific binding of ACTH to adrenocortical cells correlated very well with cAMP production but not with steroid synthesis. Corticosterone production was maximal even at the lowest concentration of ACTH used in this experiment. The concentration of ACTH required for half-maximal stimulation of cAMP production (1.76 ± 0.27 nM) was in excellent agreement with the concentration at which half-maximal binding was observed (1.96 ± 0.34 nM). When steroidogenesis and binding were measured in the concentration range of 0–60 pM (Fig. 6), a linear increase in specific binding of ^{125}I-ACTH analog to adrenocortical cells was seen with concomitant increase in steroid production. Maximal steroidogene-

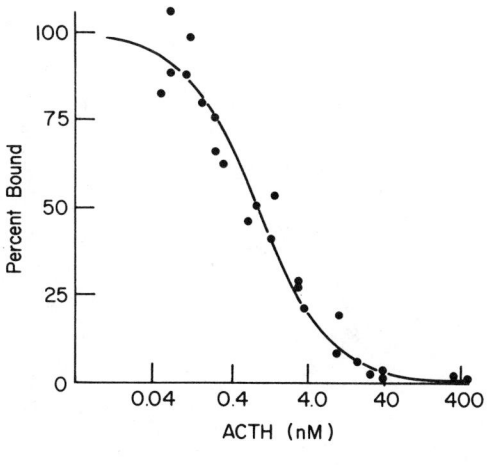
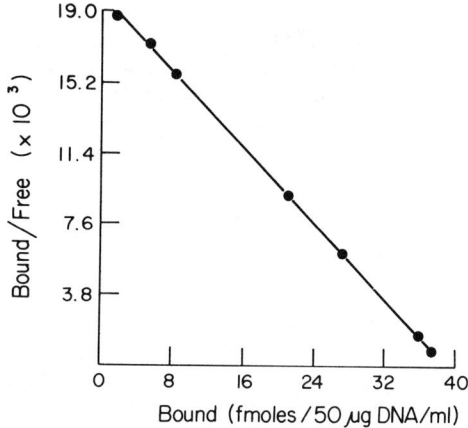

FIG. 4. Inhibition of binding ^{125}I-labeled ACTH by ACTH. (Top) Cells were preincubated with or without different concentrations of ACTH for 1 hr at 23°C and then with ^{125}I-ACTH analog (0.2 nM) for 1 hr at 23°C. The points are means of duplicate analyses of duplicate incubations derived from three separate experiments. The curve represents the best fit derived from nonlinear least squares analysis. (Bottom) Scatchard plot of the data from one representative experiment. The data from the binding inhibition curve were converted to give the amount of ^{125}I-ACTH analog specifically bound at each concentration of peptide. (From Buckley and Ramachandran, 1981.)

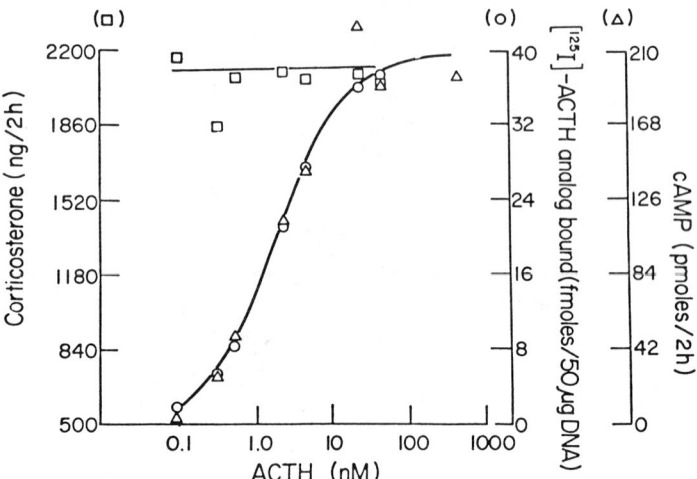

FIG. 5. Correlation of binding, cAMP production, and steroidogenesis. Cells were incubated with various concentrations of ACTH for 1 hr at 23°C and then in the presence of ^{125}I-ACTH analog for an additional hour. Aliquots of the medium were assayed for cAMP and corticosterone by specific radioimmunoassays. Nonspecific binding measured in the presence of 444 nM ACTH has been subtracted. (From Buckley and Ramachandran, 1981.)

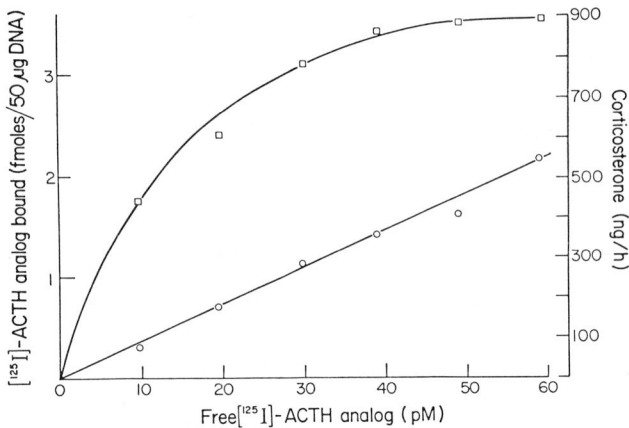

FIG. 6. Relationship between binding and steroidogenesis at low concentrations. Cells were incubated with various concentrations of ^{125}I-ACTH analog in the presence and absence of 444 nM ACTH. Binding was linear in this concentration range (correlation coefficient, 0.996). (○) Corticosterone; (□) ^{125}I-ACTH analog. (From Buckley and Ramachandran, 1981.)

sis was obtained at an ^{125}I-ACTH analog concentration of 50 pM, but there was no indication of a corresponding saturation of binding. The excellent correlation between binding of the hormone and cAMP production was further confirmed by studies with several ACTH analogs. Previous studies of the structure–activity relationships of ACTH analogs (Ramachandran et al., 1965; Ramachandran, 1973) suggested that the residues 15–18 containing the sequence Lys-Lys-Arg-Arg may play an important role in the binding of the hormone to the adrenocortical cell receptor. ACTH analogs in which the charge profile of residues 15–18 was altered in a systematic manner from +4 to +2 were examined for their ability to inhibit the binding of ^{125}I-ACTH analog to adrenocortical cells. The results in Fig. 7 (top) show that the ability of the analogs to compete for the ACTH receptor is proportional to the positive charge associated with segment 15–18. ACTH (1–19 NH$_2$), which has a net charge of +4 in the peptide segment 15–18, was more effective than ACTH (1–17 NH$_2$) which has a net charge of +3. Similarly, ACTH (1–19), which has a net charge of +3 owing to the presence of a free carboxyl group at residue 19, was more potent than ACTH (1–17) with a net charge of +2. The concentration–response curves of these peptides for cAMP production (Fig. 7, bottom) are superimposable on the binding inhibition curves. There is excellent agreement between inhibition of binding and ability to generate cAMP for these analogs not only in the ranked order of the peptides but also in the absolute values.

These results strongly support the receptor-reserve model for which indirect evidence has also been obtained from other studies. Hornsby and Gill (1978) found that bovine adrenal cells in culture progressively lost the ability to generate cAMP in response to ACTH. The decreased response was a function of cell generation and was the result of a decrease in the maximal rate of cAMP formation. The concentration of ACTH required for half-maximal stimulation of cAMP was not altered. As the cAMP response to ACTH declined in these cells, there was a progressive shift of the concentration–response curve for ACTH-induced steroidogenesis to the right. The responsiveness of these cells to prostaglandins, which stimulate cAMP production and steroidogenesis, was unaltered through the life span of the culture system. These results are best explained by the receptor-reserve concept in which only a small fraction of the cAMP formed in response to the hormone is needed for steroid synthesis.

The work of Rae et al. (1979) on genetic variants of the Y-1 mouse adrenal tumor cell provides strong evidence that cAMP and cAMP-dependent protein kinase are obligatory components of ACTH-induced steroidogenesis. In a series of mutants with an intact ACTH-sensitive adenylate cyclase system but defective cAMP-dependent protein kinases, the

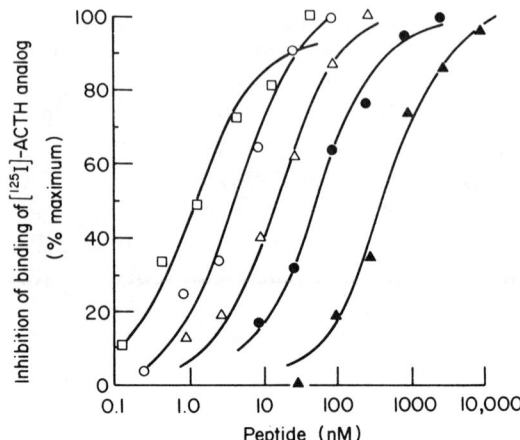
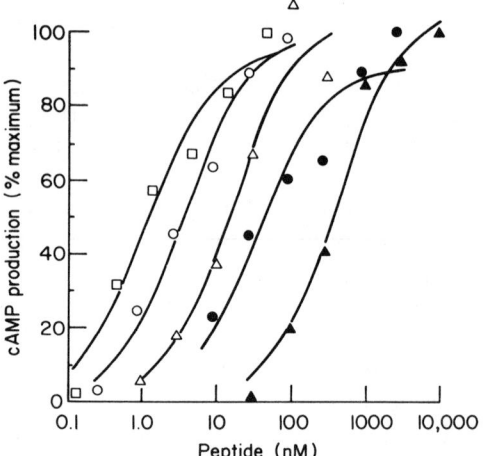

FIG. 7. Correlation of binding affinity (top) with cAMP production (bottom) for a series of ACTH analogs. Cells were incubated with ^{125}I-ACTH analog (0.2 nM) in the absence and presence of various concentrations of the peptides for 1 hr at 23°C. Points are the means of duplicate analyses of duplicate incubations, and the curves were generated by nonlinear least squares analysis. (□) ACTH; (○) ACTH (1–19 NH$_2$); (△) ACTH (1–17 NH$_2$); (●) ACTH (1–19); (▲) ACTH (1–17). (From Buckley and Ramachandran, 1981.)

abilities of ACTH or cAMP to stimulate steroidogenesis were impaired to the same extent. Together with the results obtained in the binding studies using the biologically fully potent ^{125}I-ACTH analog, all the available data suggest that ACTH interacts with a limited number of a single class of receptors on adrenocortical cells to stimulate steroidogenesis by mechanisms involving cAMP-dependent protein kinase.

C. Role of Calcium

The role of Ca^{2+} in the actions of ACTH has been the subject of numerous investigations over the past 25 years (Halkerston, 1975; Tait *et al.*, 1980). Haksar and Peron (1973) found that the requirement for Ca^{2+} in ACTH action on isolated rat adrenocortical cells was greater for events preceding the formation of cAMP than for those that follow. However, because of the studies of Lefkowitz *et al.* (1970a) it was generally accepted that Ca^{2+} was not required for the interaction of ACTH with the adrenal receptors. All previous studies attempted to elucidate the role of Ca^{2+} in ACTH binding utilized ^{125}I-labeled ACTH preparations of low biological potencies resulting from the preferential introduction of the bulky iodine atom into the tyrosine residue in position 2 of the ACTH amino acid sequence. As discussed in Section II,B, such radiolabeled preparations were detecting mostly low-affinity sites of high binding capacity, not the physiologically relevant, high-affinity, low-capacity receptors. The role of Ca^{2+} in the actions of ACTH, including its interaction with the receptor, has recently been reinvestigated using the biologically fully potent radioligand, ^{125}I-ACTH analog (Cheitlin *et al.*, 1985).

These studies show that there is an absolute requirement for extracellular Ca^{2+} for both the binding of ^{125}I-ACTH analog and for stimulation of corticosterone production. EGTA abolished both binding and steroidogenesis, and addition of Ca^{2+} restored both features in parallel (Fig. 8). Furthermore, extracellular Ca^{2+} was found to be essential for the continued occupancy of the receptor by ACTH. Removal of Ca^{2+} from the medium dramatically accelerated the dissociation of ACTH from the receptor (Fig. 9). Half-maximal dissociation of ^{125}I-ACTH analog from adrenocortical cells occurred in 32 minutes in the presence of Ca^{2+} and 3.5 min in the absence of the ion.

The absolute requirement of extracellular Ca^{2+} for the binding and continued occupancy of the receptor makes it difficult to study the effects of the ion on events subsequent to the hormone–receptor interaction. This problem was solved by covalent attachment of ACTH to its receptor by photoaffinity labeling. Ramachandran *et al.* (1981) showed that photolysis of rat adrenocortical cells in the presence of the photoreactive derivative,

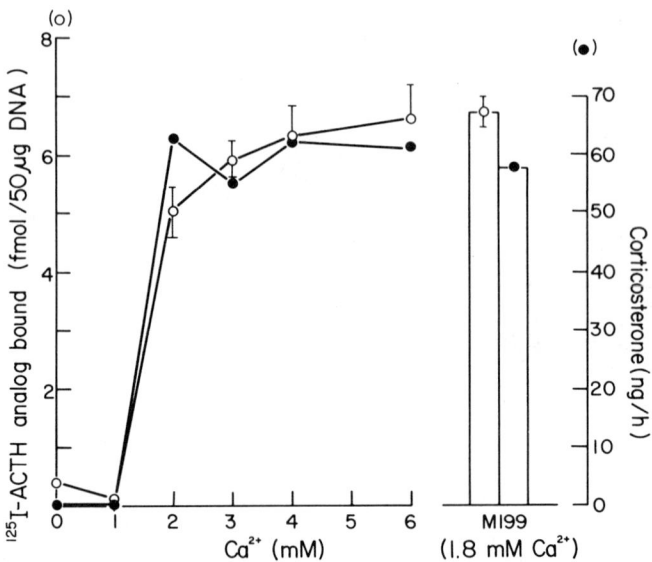

FIG. 8. Effect of Ca^{2+} on the binding of ^{125}I-ACTH analog and stimulation of steroid synthesis in adrenocortical cells. Adrenocortical cells were incubated in medium 199/0.5% BSA/0.01% bacitracin with 200 pM ^{125}I-ACTH analog in the absence or presence of 3 mM EGTA containing various concentrations of Ca^{2+} (0–6 mM), and cell bound radioactivity and corticosterone production are assessed. Nonspecific binding measured in the presence of 440 nM ACTH has been subtracted from the values which are means ± SEM of duplicate analyses of duplicate incubations. (●) Corticosterone; (○) ^{125}I-ACTH analog. (From Cheitlin et al., 1985.)

(2-nitro-5-azidophenylsulfenyl)-Trp^9-ACTH, results in the persistent activation of both corticosterone and cAMP production. This activation persisted for several hours and could not be abolished by extensive washing with anti-ACTH antibodies.

No covalent attachment of ACTH to the receptor occurred when photolysis was conducted in the presence of EGTA, supporting the conclusion derived from the binding studies that Ca^{2+} is necessary for the interaction of the hormone with the adrenocortical cell receptor. Once the hormone was covalently linked to the receptor in the presence of Ca^{2+}, the role of the ion in postbinding events could be studied. Although persistent activation of steroidogenesis induced by photoaffinity labeling was suppressed in the presence of EGTA, steroidogenesis continued unabated when the persistently activated cells were washed with Ca^{2+}-free medium and reincubated in Ca^{2+}-free medium. These results suggest that extracellular Ca^{2+} is not needed for steroidogenesis once the hormone is bound to the receptor and is able to maintain continued occupancy due to

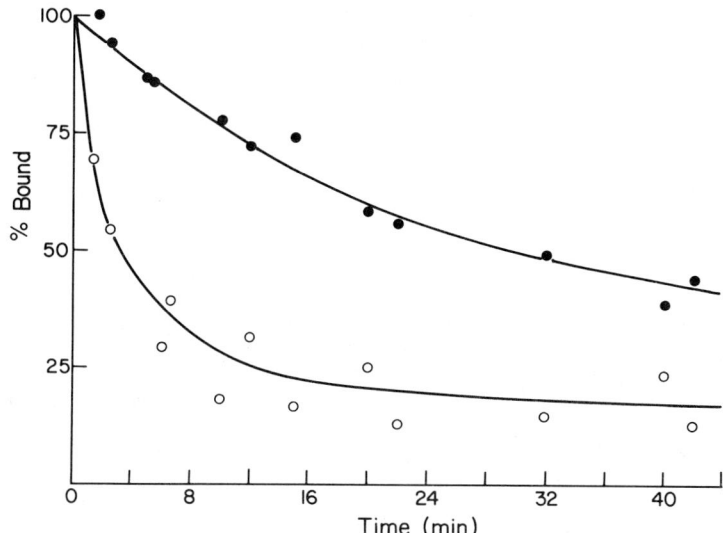

FIG. 9. Effect of Ca^{2+} on the dissociation of ^{125}I-ACTH analog from adrenocortical receptors. Cells were incubated with 200 pM ^{125}I-ACTH analog for 1 hr at 24°C and then centrifuged at 700 g for 7 minutes. The supernatant was removed and the cells were resuspended in fresh medium, divided into two parts, and incubated in the absence and presence of 3mM EGTA. Aliquots were removed in triplicate at various times and processed. Nonspecific binding measured in the presence of 440 nM ACTH has been subtracted. Points (●, ○) are means from two separate experiments. (From Cheitlin et al., 1985.)

covalent attachment. Influx of Ca^{2+} from the extracellular medium under the influence of ACTH has been considered as a possible mechanism involved in steroidogenesis (Leier and Jungmann, 1973). These results clearly show that such influx of Ca^{2+} from the extracellular fluid is not necessary for ACTH-induced steroidogenesis.

Minute quantities of Ca^{2+} bound to the membrane may, of course, play an important role in the events beyond the binding of the hormone to its receptor. Depletion of Ca^{2+} by incubation with excess EGTA is clearly deleterious and causes reversible cessation of steroidogenesis even when the hormone is covalently attached to the receptor. It is possible that Ca^{2+} is needed to maintain the hormone in a conformation favorable for productive interaction with the receptor even when covalently bound to its receptor. It is also conceivable that translocation of Ca^{2+} bound to the plasma membrane, or other organelles of the adrenocortical cell to another compartment in response to the interaction of ACTH with its receptor, may play a role in the steroidogenic action of the hormone. Studies using inhibitors of the actions of calmodulin have shown that Ca^{2+} plays an important regulatory role at intracellular sites (Carsia *et al.*, 1982). It is clear, however, that the primary and exclusive role of extracellular Ca^{2+} in the action of ACTH is to facilitate the binding of the hormone to its receptor.

IV. ACTH Receptors in Human Adrenocortical Cells

A. Fetal Adrenocortical Cells

The human fetal adrenal gland is composed of two morphologically and functionally distinct zones. The outer subcapsular zone, comprising 20% of the gland, is the precursor of the three zones of the adult adrenal cortex and synthesizes the same steroids, cortisol being the major component. The inner fetal zone constitutes the bulk of the fetal adrenal cortex and produces principally dehydroepiandrosterone (DHA) and its sulfate (DHAS). After birth the fetal zone regresses completely. Fetal and definitive zones have been separated, and the properties of cells derived from the two zones have been studies in superfusion systems and monolayer cultures (Jaffe *et al.*, 1981; Simonian and Gill, 1981). These studies indicated that ACTH induces the activity of 3β-hydroxy-steroid dehydrogenase, $\Delta^{4,5}$-isomerase, in fetal zone cultures and, thus, the steroidogenic pathway characteristic of the definitive zone and the adult adrenal cortex. Crickard *et al.* (1982, and unpublished data) investigated ACTH receptors in the two zones of the fetal adrenal cortex using ^{125}I-ACTH analog as the radioligand. These studies showed that there are specific receptors for ACTH in both zones and the affinities of the receptors in the two zones are the same (Fig. 10). However, there was a highly significant difference in the binding capacities of the two zones. Fetal zone cells derived from a fetus with a gestational age of 20 weeks bound 138 ± 12 fmol ACTH/50 μg DNA compared to 35.4 ± 9.9 fmol ACTH/50 μg DNA in the case of definitive zone cells from the same fetus. Thus, there appears to be a 4-fold higher binding capacity in the fetal zone cells which are also much larger in size compared to the definitive zone cells.

B. Adult Adrenocortical Cells

Previous investigations of ACTH receptors in normal and tumor human adrenocortical tissue (Saez *et al.*, 1975; Nishisato, 1980) employed ^{125}I-labeled derivatives of ACTH (1–24) and porcine ACTH. Saez *et al.* (1975) reported that 3.2×10^{-7} M ACTH (1–24) was required for 50% inhibition of the binding of ^{125}I-labeled ACTH (1–24) in normal human adrenal mem-

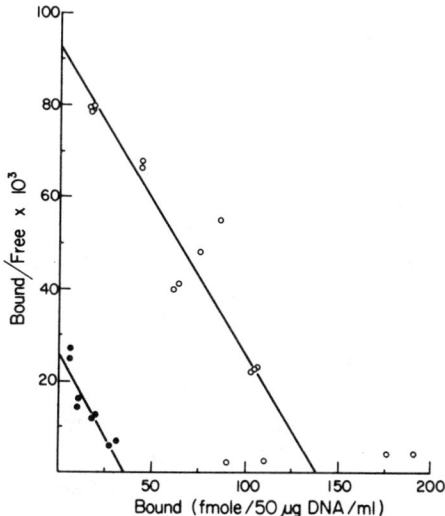

FIG. 10. Scatchard analysis of the binding of ACTH to fetal (○) and definitive (●) zone cells. (From Crickard et al., 1982, and unpublished data.)

branes. Nishisato (1980) found an apparent K_d ranging from 3.1×10^{-7} to 5.1×10^{-6} M for the binding of ^{125}I-labeled ACTH of porcine origin to crude membrane preparations of human adrenocortical tumors. These values are similar to the apparent K_d values observed for the low-affinity, high-capacity binding sites in rat and bovine adrenocortical membranes by several investigators reviewed in Section II. Since isolated human adrenocortical cells respond to physiological concentrations of ACTH in terms of steroidogenesis and cAMP production (Kolanowski and Crabbe, 1976), it is highly unlikely that the low-affinity sites are the relevant ACTH receptors. Recently, ^{125}I-ACTH analog was used to characterize ACTH receptors in human adrenocortical cells derived from adrenals obtained from brain-dead patients at the time of renal harvest (Catalano et al., 1986). ^{125}I-ACTH analog, which is equipotent with ACTH, binds to isolated human adrenocortical cells rapidly at room temperature in a reversible and saturable manner. Analysis of the binding data obtained at steady state revealed that the human adrenocortical cell ACTH receptor is remarkably similar to that of the rat adrenocortical cell in terms of affinity and capacity.

However, there are significant differences between human and rat adrenocortical cells in the coupling of the receptor to adenylate cyclase and steroidogenesis. In the rat, adrenocortical cell binding of the hormone and cAMP generation show a very high degree of correlation. The concentrations of ACTH required for half-maximal stimulation of cAMP production and the apparent K_d of the binding interaction are identical, but the concentration of the ACTH required for half-maximal stimulation of steroidogenesis is approximately 40-fold lower. In the human adrenocortical cell both cAMP production and steroidogenesis are maximally stimulated by lower concentrations of the hormone than those needed for maximal binding (Fig. 11). The concentration of ACTH causing half-maximal cAMP production is 20-fold less than that causing half-maximal inhibition of binding. Since the concentration of ACTH needed for steroidogenesis in the human adrenocortical cell is also 36-fold lower than that needed for half-maximal cAMP production, this results in a 720-fold difference in the concentration of the hormone needed for half-maximal binding and half-maximal steroid production. Thus, the human adrenocortical cell is approximately 20 times more sensitive to ACTH than the rat adrenocortical cell. Kolanowski and Crabbe (1976) investigated cortisol and cAMP production in adrenocortical cells isolated from fragments obtained at surgery for pheochromocytoma in two patients, and in one case of aldosterone-secreting adenoma, and reported that ACTH [1–24] caused half-maximal steroidogenesis and cAMP formation at concentrations of 3–8 and 120–140 pM, respectively.

Since the affinity as well as the capacity of the human adrenocortical cell ACTH receptor are virtually identical to that of the rat adrenocortical cell, the increased sensitivity of the human adrenocortical adenylate cyclase must be due to more efficient coupling of the receptor to adenylate cyclase. This could arise from intrinsic structural differences between the

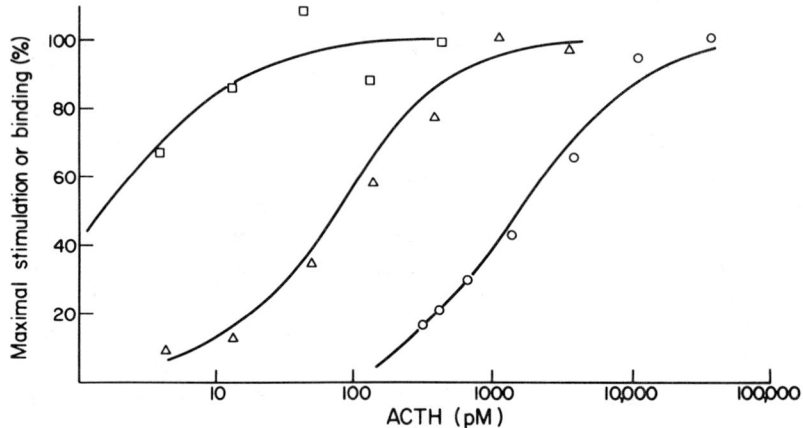

FIG. 11. Correlation of binding (○), cAMP production (△), and steroidogenesis (□) in human adrenocortical cells. (From Catalano *et al.*, 1985.)

rat and human guanine nucleotide binding proteins (G) (Gilman, 1984) or from altered concentrations of the stimulatory (Gs) and inhibitory (Gi) guanine nucleotide binding proteins in the adrenocortical cells of the two species. Although there are differences between the human and rat adrenocortical cells in the sensitivity of the adenylate cyclase, the relationship of steroidogenesis and cAMP production in the two species appears to be the same. In both species, the concentration of hormone needed for half-maximal steroidogenesis is about 40-fold lower than that needed for half-maximal cAMP production.

The human and rat adrenocortical cells are also highly similar in their requirement of calcium for the binding of the hormone to the cells. Catalano et al. (1986) found that both specific binding of ^{125}I-ACTH analog to human adrenocortical cells and ACTH-induced cAMP production were completely abolished in the presence of 3 mM EGTA. This finding supports the conclusion that the ACTH receptors in human and rat adrenocortical cells are structurally very similar, if not identical.

V. ACTH Receptors in Rat Adipocytes

The stimulation of lipolysis in rat adipose tissue by ACTH was first demonstrated by White and Engel (1958). The development of a method for preparation of isolated adipocytes by Rodbell (1964) greatly facilitated investigations of the regulation of adipocyte function by hormones. Lang et al. (1974) employed Phe^2,4,5-dehydro[4,5-^{3}H]norvaline^4 ACTH (1–24) to investigate ACTH receptors in isolated rat adipocytes. This radioligand had a specific radioactivity of 7.42 Ci/mmol and a biological potency of 10% of the hormone. Lang et al. (1974) estimated that the rat adipocyte contains approximately two million binding sites for ACTH per cell. Behrens and Ramachandran (1981) used [3,5-^{3}H]Tyr^{2,2,3}-ACTH (1–39) with a specific radioactivity of 90 Ci/mmol and full biological potency to study the regulation of ACTH receptors in rat adipocytes by glucocorticoids. These studies revealed a close parallelism between binding of ACTH and stimulation of lipolysis. No differences were found in the binding of ACTH to adipocytes derived from normal, adrenalectomized, and adrenalectomized, dexamethasone-treated rats. However, the specific radioactivity of even this initiated ACTH was inadequate for investigating the kinetic properties of specific receptors on adipocytes. It was only by the use of the ^{125}I-ACTH analog that characterization of ACTH receptors in normal rat adipocytes could be accomplished (Oelofsen and Ramachandran, 1983).

Binding of ^{125}I-ACTH analog to rat adipocytes isolated from 42-day-old rats was studied in the presence of 1 or 4% bovine serum albumin (BSA). As in the case of the adrenocortical cell, binding was highly specific, rapid, saturable, and reversible. The specific binding was maximal after 60 min at room temperature and remained stable for at least 90 min. Dissociation of the specifically bound ^{125}I-ACTH analog from adipocytes ($t_{1/2}$ = 120 min) was found to be significantly slower than in the case of the adrenocortical cells ($t_{1/2}$ = 40 min). Scatchard analysis of the binding data obtained using ^{125}I-ACTH analog in medium containing 1% BSA revealed a single class of binding sites with an apparent K_d of 170 ± 12 pM (Oelofsen and Ramachandran, 1983). Competition experiments with unlabeled ACTH also yielded a comparable value for the apparent K_d (143 ± 17 pM). The number of sites per adipocyte was quite low (521–841/cell). The stimulation of lipolysis by ACTH closely correlated with the binding (Fig. 12). The apparent K_m was 145–177 pM. Thus, in the presence of 1% BSA there appeared to be no receptor reserve in adipocytes. In this respect, the rat adipocyte differs from the rat adrenocortical cell where it is clear that stimulation of less than 3% of the receptors is sufficient for eliciting maximal steroidogenesis (Buckley and Ramachandran, 1981).

In the presence of 4% BSA (which is the usual concentration of BSA employed in metabolic studies of adipocytes), the binding curve was shifted significantly to the right (apparent K_d = 446 ± 77 pM) and the binding capacity was also significantly enhanced (1663 ± 208/cell). How-

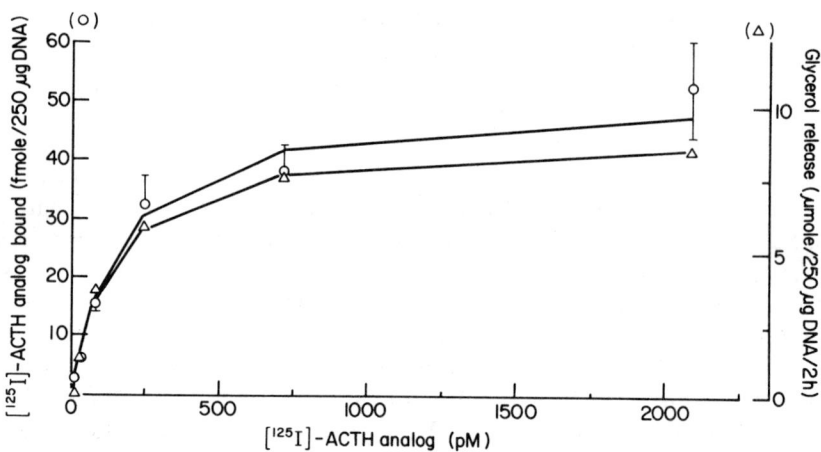

FIG. 12. Correlation of binding of ^{125}I-ACTH analog with lipolysis in rat adipocytes. Cells were incubated with several concentrations of ^{125}I-ACTH analog for 90 min at 24°C and processed as described by Oelofsen and Ramachandran (1983). The lines represent the best fit derived from weighted nonlinear least squares analysis. (△) Glycerol release; (○) ^{125}I-ACTH analog.

ever, the apparent K_m for glycerol release was unaltered (189 ± 7 pM). The highly significant increase in the apparent K_d and the binding capacity in the presence of 4% BSA may result from interactions of the adipocyte plasma membrane with free fatty acids or BSA (Brandes et al., 1982) or both. Free fatty acids have been shown to alter a variety of membrane-mediated functions including surface receptor capping in lymphocytes (Klausner et al., 1980). It is possible that at the lower BSA concentration more free fatty acids are available for intercalating into the adipocyte membrane, thereby altering the fluidity of the lipid domain of the plasma membrane.

In the presence of 4% BSA, the maximal rate of glycerol release induced by ACTH was significantly higher than in the presence of 1% BSA, but the concentration of hormone needed for half-maximal stimulation of lipolysis was comparable to that found in the presence of 1% BSA. The higher maximal rate of lipolysis in 4% BSA is probably due to more efficient trapping of free fatty acids in this incubation medium. The apparent dissociation of the binding of ACTH from the lipolytic response can be explained by the increase in the number of receptors observed in 4% BSA. It can be estimated that approximately 300 sites/cell must be occupied on the average by ACTH when half-maximal stimulation of lipolysis is produced in 1% BSA. In the presence of 4% BSA, 300 sites amount to approximately one-fifth the total number of sites and stimulation of this fraction of the receptors is produced by a concentration of ACTH (189 ± 71 pM), which is significantly lower than the apparent K_d observed in 4% BSA (446 ± 77 pM). In this medium the number of ACTH receptors is in excess of that required for maximal stimulation of lipolysis and leads to the receptor reserve phenomenon.

It is of interest to compare the ACTH binding characteristics of the adipocyte with that of the adrenocortical cell in the rat. In both cell types there appears to be only one class of receptor. The apparent K_d of the adipocyte ACTH receptor is approximately one-tenth to one-third of the apparent K_d of the adrenocortical cell receptor. However, owing to the presence of receptors much in excess of that needed for steroidogenesis, half-maximal steroidogenesis is produced by 38 ± 7 pM ACTH, a concentration well below that needed for half-maximal stimulation of lipolysis (145 ± 4.58 pM). The concentration of ACTH in rat plasma ranges in the resting state is between 5 and 14 pM and increases up to 424 pM under conditions of stress. Hence, under physiological conditions the primary action of ACTH is seen on the adrenal cortex, and only under severe stress may the lipolytic action of the hormone become an important function.

Although the rat adipocyte receptor appears to have a higher affinity than the adrenocortical cell receptor, it is apparent that the affinity is strongly influenced by the experimental conditions. It is likely that the apparent differences in properties of the adipocyte and adrenocortical cell receptor are due to the microenvironment of the receptors in the two cell types rather than to any intrinsic structural differences. Further work is needed to clarify the basis of the observed properties of ACTH receptors in these cells.

VI. ACTH Receptors in 3T3-L1 Cells

The 3T3-L1 cell line, derived from Swiss mouse embryo 3T3 cells, differentiates in culture into a phenotype that possesses the characteristics of mouse adipocytes (Green and Kehinde, 1974). These cells have been used to study the development of hormonal responsiveness and to define the role of hormone receptors and coupling factors in that process (Rubin et al., 1977, 1978; Karlsson et al., 1979; Lai et al., 1981). Catecholamines can stimulate adenylate cyclase, and insulin can stimulate glucose metabolism in the undifferentiated fibroblast cells. Differentiation of these cells into adipocytes is accompanied by an increase in both the maximal responsiveness and the sensitivity of the cells to these hormones. Receptors for these hormones are present in fibroblasts and may increase with differentiation to fat cells. In contrast, ACTH stimulates adenylate cyclase in the mature fat cells only, and the enzyme is totally insensitive to the hormone in fibroblasts (Rubin et al., 1977).

Binding of ^{125}I-ACTH analog to 3T3-L1 adipocytes and fibroblasts was investigated by Grunfeld et al. (1985). Time-dependent binding to the differentiated cells, which was inhibited by saturating concentrations of unlabeled ACTH, was demonstrated. The half-maximal concentration of ACTH for inhibition of binding was 4.3 nM compared to a value of 4 nM for half-maximal stimulation of cAMP production in the adipocytes. Scatchard analysis indicated that the 3T3-L1 adipocytes contained a single class of approximately 3500 receptors/cell. Binding of ^{125}I-ACTH analog to the adipocytes was highly specific in that it could be displaced by ACTH and its analogs but not by high concentrations of insulin, β-endorphin, or polylysine. There was an excellent correlation between the ability of the ACTH analogs to inhibit ^{125}I-ACTH analog binding to the adipocytes and the potency of the analogs in stimulating cAMP production.

In contrast, no specific binding could be demonstrated when undifferentiated 3T3-L1 fibroblasts, which are not responsive to ACTH, were studied. Although it is not possible to definitely rule out a very small number of ACTH receptors on the undifferentiated fibroblasts, Grunfeld *et al.* (1985) estimate that there must be less than 1% of receptors in the fibroblast compared to the adipocyte. These results suggest that the induction of responsiveness to ACTH accompanying differentiation of 3T3-L1 cells is accounted for primarily by the appearance of ACTH receptors.

References

Beall, J. R., and Sayers, G. (1972). *Arch. Biochem. Biophys.* **148,** 70–76.
Behrens, C. M., and Ramachandran, J. (1981). *Biochim. Biophys. Acta* **672,** 268–279.
Brandes, R., Ockner, R. K., Weigner, R. A., and Lysenko, N. (1982). *Biochem. Biophys. Res. Commun.* **105,** 821–827.
Brundish, D. E., and Wade, R. (1973). *J. Chem. Soc. Perkin Trans.* **1,** 2875–2880.
Buckley, D. I., and Ramachandran, J. (1981). *Proc. Natl. Acad. Sci. U.S.A.* **78,** 7431–7435.
Buckley, D. I., Yamashiro, D., and Ramachandran, J. (1981a). *Endocrinology* **109,** 5–9.
Buckley, D. I., Hagman, J., and Ramachandran, J. (1981b). *Endocrinology* **109,** 10–16.
Carsia, R. V., Moyle, W. R., Wolff, D. J., and Malamed, S. (1982). *Endocrinology* **111,** 1456–1461.
Catalano, R. D., Stuve, L., and Ramachandran, J. (1984). *Fed. Proc., Fed. Am. Soc. Exp. Biol.* **43,** 3037 (Abstr.).
Catalano, R. D., Stuve, L., and Ramachandran, J. (1986). *J. Clin. Endocrinol. Metab.* **62,** 300–304.
Cheitlin, R., Buckley, D. I., and Ramachandran, J. (1985). *J. Biol. Chem.* **260,** 5323–5327.
Crickard, K., Buckley, D. I., Ramachandran, J., and Jaffe, R. B. (1982). *Annu. Meet. Endocrine Soc., 64th, San Francisco, June 16–18.* p. 217, Abstr. 551.
Dedman, M. L., Farmer, T. H., and Morris, C. J. O. R. (1961). *Biochem. J.* **78,** 348–353.
Durand, P. H., and Locatelli, A. (1980). *Biochem. Biophys. Res. Commun.* **96,** 447–456.
Finn, F. M., Widnell, C. C., and Hofmann, K. (1972). *J. Biol. Chem.* **247,** 5695–5702.
Gilman, A. G. (1984). *Cell* **36,** 577–579.
Green, H., and Kehinde, O. (1974). *Cell* **1,** 113–119.
Greenwood, F. C., Hunter, W. M., and Glover, J. S. (1963). *Biochem. J.* **89,** 114.
Grunfeld, C., Hagman, J., Sabin, E., Buckley, D. I., Jones, D. S., and Ramachandran, J. (1985). *Endocrinology* **116,** 113–117.
Haksar, A., and Peron, F. G. (1973). *Biochim. Biophys. Acta* **313,** 363–371.
Halkerston, I. D. F. (1975). *Adv. Cyclic Nucleotide Res.* **6,** 99–136.
Harrington, C. A., Fenimore, D. C., and Farmer, R. W. (1978). *Biochem. Biophys. Res. Commun.* **85,** 55–61.
Haynes, R. C., and Berthet, L. (1957). *J. Biol. Chem.* **225,** 115–124.
Hofmann, K., Wingender, W., and Finn, F. M. (1970). *Proc. Natl. Acad. Sci. U.S.A.* **67,** 829–836.
Hornsby, P.-J., and Gill, G. N. (1978). *Endocrinology* **102,** 926–936.
Jaffe, R. B., Seron-Ferre, M., Crickard, K., Koritnik, D., Mitchell, B. F., and Huhtaniemi, I. T. (1981). *Recent Prog. Horm. Res.* **37,** 41–97.
Karlsson, F. A., Grunfeld, C., Kahn, C. R., and Roth, J. (1979). *Endocrinology* **104,** 1383–1390.

Klausner, R. D., Kleinfeld, A., Hoover, R., and Konnovsky (1980). *J. Biol. Chem.* **255**, 1286–1295.
Kolanowski, J., and Crabbe, J. (1976). *Mol. Cell. Endocrinol.* **5**, 255–267.
Lai, E., Rosen, O. M., and Rubin, C. S. (1981). *J. Biol. Chem.* **256**, 12866–12872.
Landon, J., Livanou, T., and Greenwood, F. C. (1967). *Biochem. J.* **105**, 1075–1080.
Lang, U., Karlaganis, G., Vogel, R., and Schwyzer, R. (1974). *Biochemistry* **13**, 2626–2633.
Lefkowitz, R. J., Roth, J., and Pastan, I. (1970a). *Nature (London)* **228**, 864–866.
Lefkowitz, R. J., Roth, J., Pricer, W., and Pastan, I. (1970b). *Proc. Natl. Acad. Sci. U.S.A.* **65**, 745–752.
Lefkowitz, R. J., Roth, J., and Pastan, I. (1971). *Ann. N.Y. Acad. Sci.* **185**, 195–209.
Leier, D. J., and Jungmann, R. A. (1973). *Biochim. Biophys. Acta* **329**, 196–210.
Lemaire, S., Yamashiro, D., Behrens, C., and Li, C. H. (1977). *J. Am. Chem. Soc.* **99**, 1577–1580.
Lowry, P. J., McMartin, C., and Peten, J. (1973). *J. Endocrinol.* **59**, 43–55.
Mackie, C., Richardson, M. C., and Schulster, D. (1972). *FEBS Lett.* **23**, 345–348.
McIlhinney, R. A. J., and Schulster, D. (1974). *Endocrinology* **94**, 1259–1264.
McIlhinney, R. A. J., and Schulster, D. (1975). *J. Endocrinol.* **64**, 175–184.
Moyle, W. R., Kong, Y.-C., and Ramachandran, J. (1973). *J. Biol. Chem.* **248**, 2409–2417.
Nakamura, M., Ide, M., Okabayashi, J., and Tanaka, A. (1972). *Endocrinol. Jpn.* **19**, 443–448.
Neher, R., and Milani, A. (1978). *Horm. Cell Regul.* **2**, 71–73.
Nishisato, K. (1980). *Nippon Naibunpi Gakkai Zasshi* **56**, 1050–1066.
Oelofsen, W., and Ramachandran, J. (1983). *Arch. Biochem. Biophys.* **225**, 414–421.
Perchellet, J.-P., and Sharma, R. K. (1979). *Science* **203**, 1259–1261.
Perchellet, J.-P., Shanker, G., and Sharma, R. K. (1978). *Science* **199**, 311–312.
Rae, P., and Schimmer, B. P. (1974). *J. Biol. Chem.* **249**, 5649–5653.
Rae, P. A., Gutmann, N. S., Tsao, J., and Schimmer, B. P. (1979). *Proc. Natl. Acad. Sci. U.S.A.* **76**, 1896–1900.
Ramachandran, J. (1973). *Horm. Proteins Polypeptides* **2**, 1–28.
Ramachandran, J., and Behrens, C. (1977). *Biochim. Biophys. Acta* **496**, 321–328.
Ramachandran, J., and Moyle, W. R. (1977). *Proc. Int. Congr. Endocrinol. 5th* pp. 520–525.
Ramachandran, J., Chung, D., and Li, C. H. (1965). *J. Am. Chem. Soc.* **87**, 2696–2708.
Ramachandran, J., Kong, Y.-C., and Liles, S. (1976). *Acta Endocrinol.* **82**, 587–599.
Ramachandran, J., Canova-Davis, E., and Behrens, C. (1980a). In "Synthesis and Release of Adenohypophyseal Hormones" (M. Jutisz and K. W. McKerns, eds.), pp. 363–380. Plenum, New York.
Ramachandran, J., Lee, C. Y., Keri, G., and Kenez-Keri, M. (1980b). In "Polypeptide Hormones" (R. F. Beers, Jr. and E. G. Bassett, eds.), pp. 295–308. Raven, New York.
Ramachandran, J., Hagman, J., and Muramoto, K. (1981). *J. Biol. Chem.* **256**, 11424–11427.
Rees, L. H., Cook, D. M., Kendall, J. W., Allen, C. F., Kramer, R. M., Ratcliffe, J. G., and Knight, R. A. (1971). *Endocrinology* **89**, 254–262.
Rodbell, M. (1964). *J. Biol. Chem.* **239**, 375–382.
Rubin, C. S., Lai, E., and Rosen, O. M. (1977). *J. Biol. Chem.* **252**, 3554–3560.
Rubin, C. S., Hirsch, A., Fung, C., and Rosen, O. M. (1978). *J. Biol. Chem.* **253**, 7570–7575.
Rubin, R. P., and Laychock, S. G. (1978). *Ann. N.Y. Acad. Sci.* **307**, 377–390.
Saez, J. M., Morera, A. M., Dazord, A., and Bataille, P. (1974). *J. Steroid Biochem.* **5**, 925–933.
Saez, J. M., Dazord, A., and Gallet, D. (1975). *J. Clin. Invest.* **56**, 536–544.
Schwyzer, R., and Karlaganis, G. (1973). *Liebigs Ann. Chem.* **760**, 1298–1308.
Simonian, M. H., and Gill, G. N. (1981). *Endocrinology* **108**, 1769–1775.

Tait, J. F., Tait, S. A. S., and Bell, J. B. G. (1980). *Essays Biochem.* **16,** 99–174.
Taunton, O. D., Roth, J., and Pastan, I. (1967). *J. Clin. Invest.* **46,** 1122–1127.
Ways, D. K., Zimmerman, C. F., and Ontjes, D. A. (1976). *Mol. Pharmacol.* **12,** 789–799.
White, J. E., and Engel, F. L. (1958). *J. Clin Invest.* **37,** 1556–1563.
Yanagibashi, K., Kamiya, N., Ling, G., and Matsuba, M. (1978). *Endocrinol. Jpn.* **25,** 545–551.

3
Biosynthesis of ACTH and Related Peptides

EDWARD HERBERT,* MICHAEL COMB,†
GARY THOMAS,* DANE LISTON,*
OLIVIER CIVELLI,* MITCHELL MARTIN,*
AND NEAL BIRNBERG†

Department of Chemistry
University of Oregon
Eugene, Oregon 97403

I. Introduction

ACTH is a peptide hormone which is released from the anterior pituitary into the circulation in response to the presence of the hypothalamic peptide, corticotropin releasing factor (CRF). ACTH stimulates the release of glucocorticoids and mineralocorticoids from the adrenal cortex. These steroids modulate a variety of biological responses, including blood glucose utilization and production and salt balance. Glucocorticoids also inhibit further release of ACTH from the pituitary by feedback inhibition at the level of the pituitary and hypothalamus.

In the past few years ACTH and related peptides such as melanocyte stimulating hormone (α-MSH) have been detected in many different tissues including the brain. The function of these peptides in extrapituitary tissues is not known, but there is strong suggestion that some of them may play a role in learning processes in the brain (DeWied, 1982).

The purpose of this chapter is to review what we know about ACTH synthesis, primarily in the pituitary but also in extrapituitary tissues.

---

* Present address: Institute for Advanced Biomedical Research, The Oregon Health Sciences University, Portland, Oregon 97201.

† Present address: Molecular Biology Department, Massachusetts General Hospital, Boston, Massachusetts 02114.

Emphasis will be placed on the role that molecular biology has played in helping us to understand the pathway of ACTH biosynthesis and the genetic regulation of ACTH expression. We will also discuss the potential that exists for future discoveries as a result of the application of gene transfer techniques to the study of regulation of production of ACTH and related peptides.

Although ACTH was the first pituitary polypeptide hormone to be isolated and sequenced (Li et al., 1954, 1955; Howard et al., 1955), its biosynthesis presented an interesting puzzle for many years. The only polypeptide hormone whose synthesis had been studied prior to ACTH was insulin. In 1967 Steiner and co-workers demonstrated that insulin is synthesized in the form of a precursor, proinsulin, which is not quite twice the size of insulin. It appeared likely from that study that ACTH was also synthesized in a precursor form. This expectation was realized when, in the early 1970s, high-molecular-weight forms of ACTH were detected in several tissues. However, the real breakthrough in our understanding of ACTH biosynthesis did not occur until mouse pituitary tumor cells were introduced as a model for anterior pituitary corticotrophs. These cells (AtT-20-D16v cells) proved to be an excellent model system for synthesis studies because they produced large quantities of ACTH (Herbert et al., 1978). The study of ACTH biosynthesis was also influenced by the observation that ACTH and β-lipotropin (LPH) are both present in corticotrophic cells (Moon et al., 1973; Pelletier et al., 1977). This close association of the two peptides raised the possibility of their cosynthesis in the same precursor protein. This possibility was realized in 1977 when Mains and Eipper and Roberts and Herbert demonstrated that a newly synthesized protein in AtT-20 cells could be immunoprecipitated with antibodies to either ACTH or β-LPH. Mains et al. (1977) showed the existence of the protein in intact AtT-20 cells, whereas Roberts and Herbert demonstrated its production in an mRNA-dependent cell-free translation system (1977a). In the latter case, RNA was extracted from AtT-20 cells and translated in the cell-free system in the presence of radioactive amino acids. Following immunoprecipitation with either anti-ACTH or anti-β-LPH, a single radioactive protein was detected in the immunoprecipitates by SDS–gel electrophoresis (Roberts and Herbert, 1977a,b).

This protein, which had an apparent molecular weight of 28,500, was shown by tryptic and chymotryptic peptide mapping to contain sequences of both ACTH and β-LPH. Furthermore, the arrangement of ACTH and β-LPH in the precursor was established by a novel technique of gradient labeling of nascent peptides in ribosomes (Roberts and Herbert, 1977b). Two years later, the complete structure of the precursor was revealed by the use of recombinant DNA approaches (Nakanishi et al., 1979). The recombinant DNA sequencing work confirmed many features of the structure of the precursor determined by peptide mapping, including the arrangement of ACTH and β-LPH in the precursor and, in addition, revealed some new aspects of structure such as the existence of a potentially new bioactive peptide in the cryptic region (N-terminal region) of the molecule with homology to α- and β-MSH (Fig. 1). In addition, all bioactive domains were shown to be flanked by pairs of basic amino acid residues, mostly Lys-Arg sequences which are now known to represent potential proteolytic cleavage signals. Figure 1 shows how these domains are cut out of the precursor in the anterior lobe of mouse and rat pituitary (see also Fig. 5).

The structural studies outlined above show that the precursor to ACTH contains the sequences of a variety of other peptides including β-LPH. β-LPH, in turn, contains the sequences of β-endorphin and β-MSH. In addition, the N-terminal portion of the molecule contains the sequence of another peptide named γ-MSH as pointed out above. The N-terminus of the cell-free-synthesized protein has a hydrophobic signal sequence of 26 amino acids that is required for translocation of the precursor across the endoplasmic reticulum membrane (Policastro *et al.*, 1981).

The bioactivity of many of the peptides derived from the precursor to ACTH is not very well understood. ACTH itself is known to stimulate the production of steroids (glucocorticoids and mineralocorticoids) in the adrenal cortex and to suppress activity of the immune system as mentioned above. β-LPH has weak lipolytic activity, whereas β-endorphin has potent opioid activity. However, the function of circulating β-endorphin is not known. The three melanocyte-stimulating peptides in the precursor (α-, β-, and γ-MSH) are so named because of their ability to stimulate melanization in lower vertebrates (Burgers *et al.*, 1963). The γ-MSH region of the precursor is also reported to act synergistically with ACTH

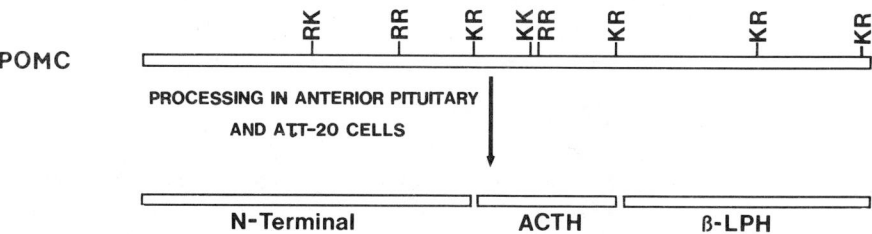

FIG. 1. Schematic presentation of the structure of POMC and the fragments produced from POMC in the anterior pituitary. The pairs of basic amino acids are indicated in the structure of POMC.

to stimulate production of steroids (Pedersen and Brownie, 1980; Seidah et al., 1981a). Thus, the precursor contains peptides that exhibit at least three different kinds of activities: steroidogenic, morphinomimmetic, and melanocyte stimulating activities. For that reason, the precursor has been named pro-opiomelanocortin, or POMC for short.

POMC is the first of a number of proteins that has been discovered in the last 10 years that contain the sequences of more than one bioactive peptide (Douglass et al., 1984). These kinds proteins have been referred to in the literature as "polyproteins" or polyfunctional proteins. In the case of POMC, it appears that peptides from different regions of the molecule may act together to coordinate components of a complex behavior pattern such as the response to stress.

II. Structure of POMC Gene and Protein in Different Species

The availability of POMC cDNA clones has allowed several laboratories to study the structure of the POMC gene in different species. The first POMC gene fragment to be described was for the human. The structure of an entire POMC gene was then determined for the cow (Nakanishi et al., 1981). The structure of the POMC gene from other species rapidly followed the characterization of the bovine gene. These include a characterization of the complete human gene (Takahashi et al., 1983), rat gene (Drouin and Goodman, 1980), and mouse gene (Uhler et al., 1983b; Notake et al., 1983). The overall structure of the POMC gene is highly conserved among these different species and is schematically diagrammed in Fig. 2. In all cases, a large 3' exon contains the nucleotides coding for all of the biologically active peptides (ACTH, LPH, MSH, and β-endorphin) and the majority of the N-terminal portion of the precursor.

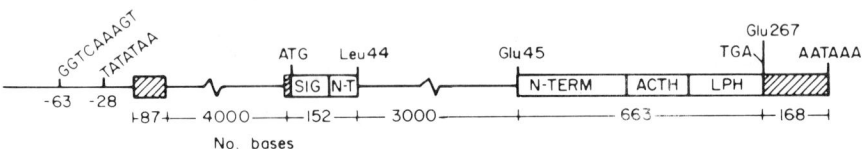

FIG. 2. Schematic diagram of the human POMC gene. Solid lines represent intronic regions while boxed areas represent exons. Nucleotide numeration starts at the cap site, amino acid numeration starts at the intitiator methionine. At bases −63 and −28 two important sequences are indicated. The number of bases of the introns are approximate. ATG indicates the start of the coding region; TGA the stop codon. AATAAA shows the poly(A) addition site found in the cDNA. SIG, signal peptide; N-T or N-TERM, N-terminal fragment; LPH, lipotropic hormone. The hatched boxes indicate the extent of the 5' and the 3' untranslated region of the mRNA.

Some 2–3 kb upstream (5') is a smaller exon which contains sequences coding for the remainder of the amino-terminal portion of the precursor molecule, the initiator methionine, and a few bases of the 5' untranslated region of the mRNA. The remaining sequences coding for the 5' untranslated region of the mRNA are found further upstream (about 4 kb) on one small exon. The POMC gene has been localized to chromosome 2 in human and chromosome 19 in mouse (Uhler *et al.*, 1983b).

Southern blotting experiments and characterization of POMC genomic clones from genomic libraries have indicated that only one POMC gene is present in the human genome. The same holds true for the number of POMC genes in the bovine and rat genomes. In contrast, two groups (Notake *et al.*, 1983; Uhler *et al.*, 1983b) have reported the presence of two POMC genes in the mouse genome, one of which is a pseudogene. The pseudogene exhibits 92% homology with 533 base pairs of the functional POMC gene, including the coding regions for ACTH and β-LPH. However, the presence of a translation termination codon in place of the codon for the first amino acid in β-endorphin and a mutation in a codon for a dibasic amino acid cleavage site within the protein predict that β-endorphin would not be present in the translation product and that ACTH would not be cleaved from the precursor by a trypsin-like cleavage enzyme. Thus, the POMC pseudogene in mouse cannot lead to expression of a protein containing β-endorphin. Finally, the mouse POMC pseudogene sequence is flanked on both sides by direct repeats 10 base pairs in length, raising the possibility that the pseudogene may have arisen via the formation of an aberrant transcript of the functional gene followed by the insertion of its cDNA copy into the mouse genome (Notake *et al.*, 1983).

Since the POMC mRNA sequences are known for five mammals [rat (Drouin and Goodman, 1980); cow (Nakanishi *et al.*, 1979); pig (Boileau *et al.*, 1983); mouse (Uhler and Herbert, 1983); human (Chang *et al.*, 1980)], an amphibian, *Xenopus laevis* (Martens *et al.*, 1985), and a fish, salmon (Soma *et al.*, 1984)], it is possible to make some interesting comparisons of structure at both the nucleic acid and protein levels. The most conserved regions of nucleotide sequence are centered around each of the three MSH units. The 5' untranslated regions show homologies near the TATA and the CAAT boxes which serve as transcriptional regulatory signals in all eukaryotic genes. Nucleotide sequence homology is also observed in a region located 200–350 bases upstream from the cap site (start of transcription). In human, this region contains two overlapping sequences that are homologous with two sequences located upstream from the transcription start site of the mouse mammary tumor virus long terminal repeat (Takahashi *et al.*, 1983). These overlapping sequences are also found upstream from the cap sites of the prolactin gene and the growth hormone gene. Since the expression of all of these genes is regulated by glucocorticoids, it has been suggested that these conserved sequences might be involved in glucocorticoid regulation.

Comparison of the amino acid sequence of mammalian POMC with that of the amphibian and fish reveals that the arrangement of bioactive domains is highly conserved among vertebrates. Pairs of basic amino acids delineate these domains in all species (Martens et al., 1985). The high degree of conservation of β-endorphin between *Xenopus* and mammals is noteworthy since the structure of this peptide is different in salmon (Martens et al., 1985). It is also very interesting to note that all four mammalian species have serine in position 1 of ACTH, whereas *Xenopus* has alanine in this position. While this substitution is not known to alter the function of ACTH, it may alter the activity of the MSH derived from ACTH (Martens et al., 1985). Finally, it is interesting to note that the spacer regions in POMC between γ-MSH and ACTH and between α-MSH and β-LPH are quite different among the six species compared, but they all have an acidic nature in common, similar to the spacer regions between enkephalin units in proenkephalin (Comb et al., 1982; Gubler et al., 1982; Noda et al., 1982) and prodynorphin (Kakidani et al., 1982; Civelli et al., 1985). The hydrophilic character of these regions in the opioid precursors might be important for efficient proteolytic processing of the precursors as pointed out previously by Comb et al. (1982).

III. Distribution and Site of Synthesis of POMC-Derived Peptides

POMC-derived peptides such as ACTH and endorphins are found predominantly in the two lobes of the pituitary gland, their principal sites of synthesis. These peptides, however, have also been reported in a wide variety of extrapituitary tissues including the pancreas, gastrointestinal tract, placenta, thyroid, mast cells, the male reproductive tract, and the central nervous system (CNS) (Nakai et al., 1978; Larsson, 1977; Krieger et al., 1977; Orwoll and Kendall, 1980; Tsong et al., 1982). In these studies only the bioactive forms of these peptides were detected, not the precursor, POMC. Thus, it is not clear from these studies if ACTH and endorphin are derived from POMC in extrapituitary tissues. The identification of larger forms of ACTH and β-endorphin, i.e., the POMC precursor, in some extrapituitary tissues suggests that POMC is synthesized and processed in these tissues (Krieger et al., 1980).

The evidence that β-endorphin, ACTH, and α-MSH are produced in the brain comes from the finding of a high-molecular-weight precursor, containing both ACTH and β-endorphin antigenic determinants, in cultured neonatal rat hypothalamic neurons.

A more definitive way of determining the site of synthesis of POMC is to test for the presence of POMC mRNA by hybridization procedures using cloned cDNA probes specific for POMC mRNA. The distribution of POMC mRNA in various rat brain tissues shows that the hypothalamus, amygdala, and cerebral cortex contain POMC transcripts (mRNA), while the cerebellum and midbrain do not (Civelli et al., 1982). In addition, POMC transcripts in the amygdala and cortex appear to be slightly smaller in size than POMC transcripts isolated from the hypothalamus. The reason for this difference in size is not known.

Two important pieces of information can be derived from these data. First, the distribution of POMC mRNA in the rat brain is an indication of the sites of synthesis of this mRNA and, presumably, of POMC as well. Second, the levels of mRNA determined by hybridization can be correlated with the levels of the bioactive peptides derived from POMC. In most cases a direct correlation is observed, confirming that the bioactive peptides previously detected in these tissues by radioimmunoassay are present as a result of direct synthesis in these tissues, and not as a result of transport from a secondary site of synthesis.

Northern blotting and solution hybridization are sufficient for quantitating the level of a particular species of mRNA in whole tissues. However, in the brain, POMC is produced in tissues made up of many types of cells and it would be advantageous to be able to determine which cells are transcribing polyprotein genes in these complex tissues. In situ hybridization histochemistry is a recently developed technique (Brahic and Haase, 1978) that is being used to make these types of determinations. In this technique, serial sections of tissue are fixed and incubated with a radiolabeled or biotinated (Langer-Safer et al., 1982) DNA or RNA probe complementary to POMC mRNA. As with Northern blotting, the resulting DNA/RNA hybrids or RNA/RNA complexes are detected by autoradiography.

Two groups (Hudson et al., 1981; Gee and Roberts, 1982) have used POMC cDNA as a hybridization probe to study the differential expression of the POMC gene in the various lobes of the rat pituitary. Using these probes, approximately 3–5% of anterior lobe cells appears to contain POMC transcripts, while greater than 90% of the intermediate lobe cells shows cytoplasmic localization of POMC mRNA. By combining immunohistochemical methods with in situ hybridization histochemistry, it has also been shown that the same cells contain both POMC peptides and

POMC mRNA (Gee and Roberts, 1982). These data suggest that the presence of the POMC peptides in these cells is due to their local synthesis and is not the result of plasma uptake from secondary sites of synthesis.

Unquestionably, *in situ* hybridization histochemistry is a valuable tool which will complement and in many instances replace other techniques currently used to localize the sites of expression of polyproteins.

IV. Regulation of Expression of POMC Genes

The studies designed to determine the structure of the POMC gene and mRNA have led to the development of POMC-specific DNA probes. These probes have been used to study the mechanisms underlying POMC gene regulation in different tissues. The next section will review some of our knowledge concerning the regulation of POMC gene expression.

A. Factors Affecting Secretion of POMC Peptides from the Pituitary

In the anterior lobe (AL), the secretion of POMC-derived peptides is positively regulated by corticotropin releasing factor (CRF) (Vale *et al.*, 1981) and negatively regulated by endogenous glucocorticoids released from the adrenal cortex (Fig. 3) (Watanabe *et al.*, 1973; Fleischer and Rawls, 1970). In the neurointermediate lobe (NIL), dopaminergic neurons originating in the hypothalamus impinge on POMC-containing cells and inhibit the release of POMC-derived peptides (Vermes *et al.*, 1980; Farah *et al.*, 1982). The administration of dopamine antagonists, such as haloperidol, thus stimulates the release of POMC peptides from the NIL (Hollt and Bergmann, 1982) and increases the levels of POMC peptides within the NIL (Hollt *et al.*, 1982; Lepine and Dupont, 1981). These compounds have no effect on the secretion of POMC peptides from anterior lobe corticotrophs.

The experiments described below demonstrate that the same factors which affect secretion of POMC peptides from the pituitary also affect the levels of POMC mRNA in a tissue-specific fashion.

B. POMC Gene Regulation in the Anterior Lobe of the Pituitary

Nakanishi *et al.* (1977) have shown that the administration of the synthetic glucocorticoid dexamethasone (DEX) to adrenalectomized rats results in a marked suppression of POMC mRNA activity in the pituitary

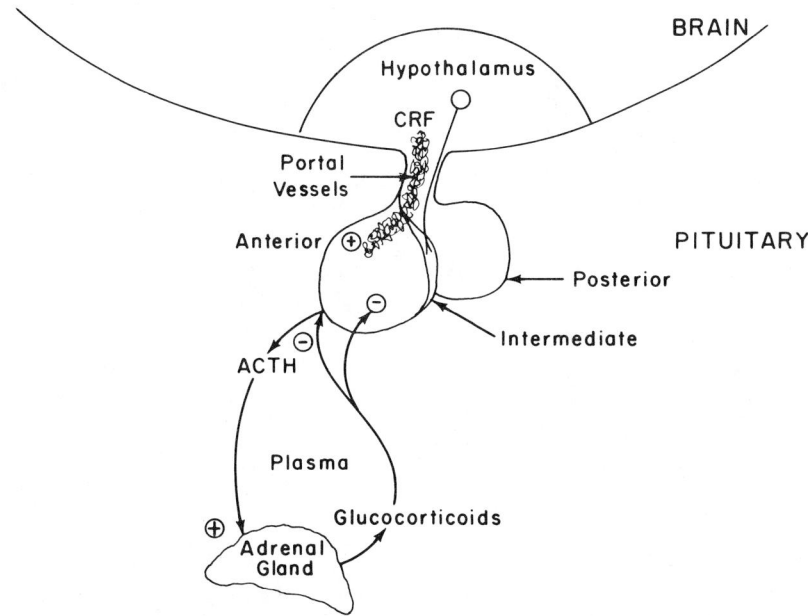

Fig. 3. Model for the regulation of ACTH production. Corticotropin releasing factor (CRF) is synthesized in the hypothalamus, discharged in portal vessels to reach the anterior lobe of the pituitary. There it stimulates POMC peptide production, in particular, ACTH. ACTH is secreted into the plasma, interacts on receptors found at the surface of the adrenal cortex, and stimulates, in particular, the production of glucocorticoids. These is turn inhibit ACTH production in the anterior pituitary lobe. They have a rapid effect on the secretion of the ACTH peptides and a long-term effect on POMC synthesis. It is also possible that the glucocorticoids act on the hypothalamus to inhibit CRF synthesis and/or release. POMC production in the intermediate lobe of the pituitary is under different control. Dopaminergic neurons originating in the hypothalamus are the major regulators of the POMC synthesis and release.

as assayed in cell-free protein synthesizing systems. Since the pituitary was not dissected into individual lobes in this study, the specific contribution of anterior lobe cells to this response could not be accurately evaluated.

The same techniques were then used to determine the effects of steroid hormones on POMC mRNA activity in mouse anterior pituitary tumor cells that secrete POMC peptides (AtT-20 cells) (Nakamura *et al.*, 1978; Roberts *et al.*, 1979). The addition of DEX to culture medium reduced the level of POMC mRNA activity to 30–40% of that in untreated cells. This reduction in POMC mRNA activity occurs without affecting the rate of processing of POMC to ACTH and β-endorphin (Roberts *et al.*, 1979).

More recently, mouse and rat POMC cDNA clones have been used as hybridization probes to accurately determine the effects of adrenalectomy and subsequent DEX administration on POMC mRNA levels in the anterior and intermediate lobes of rat pituitary (Herbert *et al.*, 1981; Birnberg *et al.*, 1983). Following adrenalectomy, POMC mRNA levels increased markedly, reaching 15- to 20-fold the control level at 18 days postoperation. When DEX was administered to rats 8 days after adrenalectomy, the above events were reversed (Birnberg *et al.*, 1983). After 5 days of treatment with DEX, POMC mRNA levels had returned to control levels. Nuclear run-off experiments (a measure of the levels of nascent POMC mRNA being synthesized in the cell nucleus) also revealed that adrenalectomy increased the rate of transcription of the POMC gene 20-fold within 1 hr after the operation (Birnberg *et al.*, 1983) (Fig. 4). As before, the administration of DEX immediately after adrenalectomy suppressed the increased rate of transcription of the POMC gene in the anterior lobe (Fig. 4). The above effects are specific to POMC gene expression in the anterior lobe; POMC mRNA levels and the rate of transcription of the POMC gene in the NIL are only slightly altered, if at all, by adrenalectomy or DEX administration (Schachter *et al.*, 1983; Birnberg *et al.*, 1983). Also, the level of POMC mRNA in the hypothalamus is not altered by the above treatments. The effects of adrenalectomy and DEX treatment on transcription of the growth hormone gene in the anterior pituitary are in the opposite direction to that of the POMC gene (Fig. 4).

In situ hybridization has been used to determine that the increase in POMC mRNA levels in the anterior lobe following adrenalectomy is due to a number of factors, including enlarged cell volume and an increase in the number of POMC-producing cells (Gee and Roberts, 1982). This technique clearly allows for a more redefined analysis of the factors involved when a heterogeneous tissue is being studied.

C. POMC Gene Regulation in the Neurointermediate Lobe of the Pituitary

Cell-free translation and immunoprecipitation (Hollt *et al.*, 1982) and RNA dot blotting (Chen *et al.*, 1983) have been used to study the effects of dopamine agonists and antagonists on POMC mRNA levels in the rat NIL. Administration of the dopamine antagonist, haloperidol, to rats results in a 4- to 6-fold (time- and dose-dependent) increase in the level of NIL POMC mRNA. This stimulatory effect is observed as early as 6 hr after administration. In contrast, ergocryptine, a dopamine agonist, de-

3. BIOSYNTHESIS OF ACTH

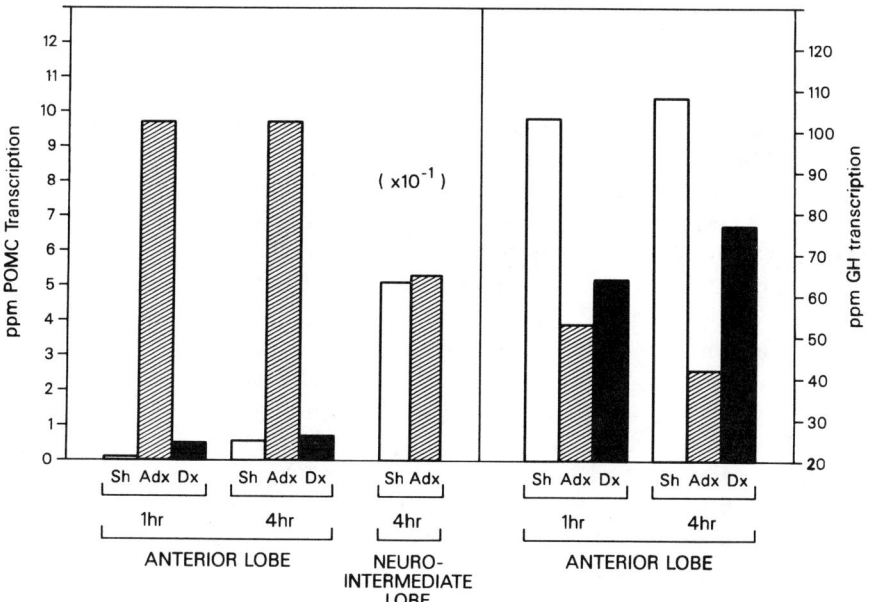

FIG. 4. Adrenalectomy causes elevation of POMC gene transcription in the anterior pituitary. Rats were subjected to bilateral adrenalectomy or sham operations. Half of the adrenalectomized animals were immediately injected with 100 μg of dexamethasone in the pericardium. At 1 and 4 hr after surgery, the animals were sacrificed, the pituitaries were removed, and the lobes were separated and frozen in liquid nitrogen. Nuclei were purified (Evans *et al.*, 1982) and assayed for POMC and growth hormone (GH) transcription rates by labeling nascent transcripts *in vitro* with [^{32}P]UTP, -GTP, and -CTP and hybridizing the radioactive RNA to plasmid DNA bound to nitrocellulose. All nondexamethasone-treated animals were injected with an equal volume of saline vehicle. DNAs from pMKSU 16 (Uhler and Herbert, 1983) and from pGH1 (Harpold *et al.*, 1978) bound to nitrocellulose filters were used as hybridization probes of specific POMC and GH transcription rates, respectively. pBR322 filters were included as controls and for background determinations. Results are reported in cpm of specific bound hybrid per 10^6 input cpm (ppm). Sh, sham operated; Adx, adrenalectomized; Dx, adrenalectomized, dexamethasone treated.

creases 2- to 3-fold the level of POMC mRNA in the rat NIL (Chen *et al.*, 1983). The time-dependent changes in POMC mRNA levels and the magnitude of these changes suggest that dopaminergic compounds modulate POMC mRNA levels in the NIL in the same fashion as they regulate POMC peptide secretion. The mechanisms underyling dopaminergic modulation of POMC mRNA levels in the NIL remain to be elucidated. It is also worthwhile to note that dopaminergic compounds have no effect on POMC mRNA levels in the anterior lobe (Chen *et al.*, 1983).

D. DIFFERENTIAL REGULATION OF THE POMC GENE

The experiments described above suggest that different factors regulate transcription of the POMC gene in the NIL and AL. Since there is only one POMC gene in the rat (Roberts *et al.*, 1982), these tissues may differ in their ability to respond to modulatory factors.

Glucocorticoids alter the level of POMC mRNA in the AL but have no effect on POMC mRNA levels in the NIL. Autoradiographic studies with labeled steroids have shown that there is no specific uptake of glucocorticoids into the nuclei of POMC-producing cells in the NIL as there is in the AL (Warembourg, 1975; Rees *et al.*, 1977). Thus, NIL cells do not appear to contain functional glucocorticoid receptors.

The experiments by Birnberg *et al.* (1983) also reveal that changes in the rate of secretion of POMC-derived peptides from the anterior lobe following adrenalectomy or DEX treatment are much more rapid than the change observed in the level of POMC mRNA. These data suggest that secretion of POMC peptides and transcription of the POMC gene are regulated independently by glucocorticoids in the anterior lobe.

POMC mRNA levels in the NIL are affected by dopaminergic compounds while little effect is observed on the maintenance of POMC mRNA levels or on transcription of the POMC gene in the AL. This tissue-specific regulation could be due to the fact that dopamine receptors are present on NIL cells but are not associated with anterior lobe corticotrophs (Gudelsky *et al.*, 1980; Nansel *et al.*, 1979).

V. Processing of POMC in the Pituitary and Brain

The peptide domains of POMC must be cleaved out of the precursor and modified in order to become bioactive. Modifications such as proteolytic cleavage, glycosylation, phosphorylation, acetylation, sulfation, and amidation frequently occur in a well-defined order as the proteins and peptides move through the endoplasmic reticulum, Golgi complex, and secretory vesicles on their way to being secreted. Amino acid sequences specify interactions involved in compartmentalization of the proteins, sites of proteolytic cleavage, and chemical modification.

Transport of secretory proteins across the membranes of the endoplasmic reticulum requires a hydrophobic N-terminal signal sequence of 25–30 amino acids (Lingappa and Blobel, 1980; Blobel and Dobberstein, 1975a,b). The signal sequence is removed from the protein prior to completion of translation (Dorner and Kemper, 1978; Shields and Blobel, 1978; Patzelt *et al.*, 1978; Jackson and Blobel, 1977). As mentioned before, Policastro et al. (1981) have shown that mouse POMC has a 26-amino acid signal sequence which is removed during biosynthesis in AtT-20 cells.

One important feature of all polypeptide precursors including POMC (Fig. 1) is that the bioactive domains are flanked by pairs of basic amino acid residues (either Lys-Arg, Lys-Lys, or Arg-Arg) as mentioned earlier. This suggests that trypsin-like enzymes are involved in an endoproteolytic cleavage reaction. The products of this cleavage are further modified by the action of a carboxypeptidase-like enzyme, which removes the C-terminal basic amino acid to produce a bioactive peptide, and possibly an amino peptidase, which removes N-terminal amino acids. In the POMC precursor, eight different pairs of basic amino acid residues can be found. Maturation of bioactive peptides implies cleavages at these sites. However, not all of these sites in POMC are cleaved in AtT-20 cells or in the anterior pituitary. For example, ACTH (1–39) contains two pairs of basic amino acids that are not cleaved in the anterior pituitary (see Fig. 1).

Maturation of POMC-derived bioactive peptides also involves amino acid residue modification. Carboxy-terminal amidation has been observed in the formation of α-MSH from POMC (Eipper et al., 1983; Bradbury et al., 1982; Scott et al., 1973, 1976). This modification is required for bioactivity of αMSH. Glycosylation of a protein at asparagine residues requires the sequence Asn-X-Thr or Asn-X-Ser (Marshall, 1974; Hubbard and Ivatt, 1981). Glycosylation can also occur at serine or threonine residues (Marshall, 1974; Hubbard and Ivatt, 1981). In rodent POMC, glycosylation occurs at two POMC sites: the γ-MSH and corticotropin-like intermediate lobe peptide (CLIP). The CLIP-related glycosylation event, however, occurs only in a portion of the POMC molecules. Other posttranslational modifications have been detected, including phosphorylation (Raese et al., 1980; Bennett et al., 1981a,b; Eipper and Mains, 1982) and acetylation (Zakarian and Smyth, 1980; Smyth et al., 1979; Eipper and Mains, 1981; Liotta et al., 1981; Seizinger and Hollt, 1980; Akil et al., 1981).

VI. Processing Pathways of POMC in the Anterior and Neurointermediate Lobes of the Pituitary and the Brain

The anterior and neurointermediate lobes of rat pituitary provide a very convenient system for studying tissue-specific expression of POMC because there are marked differences in the types of peptides derived from the ACTH and β-LPH portions of the precursor in the two lobes of the pituitary. The anterior lobe contains predominantly ACTH (1–39) while the intermediate lobe of the pituitary contains high levels of α-MSH and

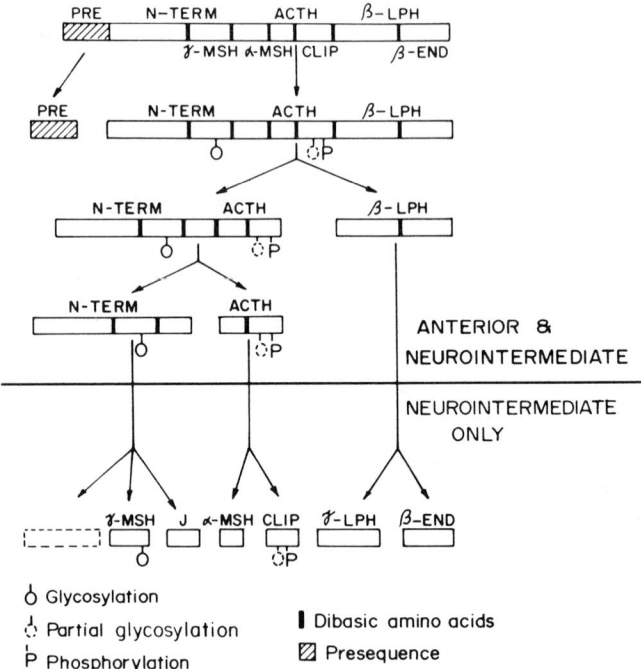

FIG. 5. Tissue-specific processing pathways of the POMC precursor. Processing occurring in the anterior and neurointermediate lobes is shown above, processing specific to the neurointermediate lobe is below. The peptides produced are shown sequentially. PRE, presignal sequence; J, joining; CLIP, corticotropin-like intermediate product; β-END, β-endorphin.

CLIP as shown in Fig. 5 (Scott et al., 1974, 1976; Roberts et al., 1978; Mains and Eipper, 1979; Eipper and Mains, 1978; Kraicer, 1977; Gianoulakis et al., 1979). Thus, Lys-Arg sites that are masked in ACTH (1–39) in the anterior lobe are accessible to cleavage enzymes in the neurointermediate lobe to yield α-MSH and CLIP. The two lobes also differ in the amounts of β-LPH, β-endorphin, and acetylated derivatives of β-endorphin that they contain. The anterior lobe contains mainly β-LPH, whereas the neurointermediate lobe contains predominantly β-endorphin and derivatives of β-endorphin (Zakarian and Smyth, 1980; Smyth et al., 1979; Eipper and Mains, 1981; Liotta et al., 1981; Seizinger and Hollt, 1980; Akil et al., 1981; Baizman and Cox, 1978; Baizman et al., 1979; Loeber et al., 1979; Lissitzky et al., 1978). Despite the differences in the ACTH/endorphin peptides in the two lobes of the pituitary, the forms of POMC appear to be identical as determined by sequencing mRNA in the two lobes (Oates and Herbert, 1984, and unpublished studies).

Pulse-label and pulse-chase studies with mouse and rat pituitary cell cultures show that the early glycosylation and cleavage steps in the processing of POMC are the same in the two lobes of rodent pituitary as shown in Fig. 5 (Roberts *et al.*, 1978; Eipper and Mains, 1978; Rosa *et al.*, 1980; Hinman and Herbert, 1980). Glycosylation of the N-terminal portion of POMC occurs first in the γ-MSH region of the molecule. Approximately half of the POMC molecules are also glycosylated in the ACTH portion (at Asn residue 29 in ACTH). The latter glycosylation site is absent in bovine (Li *et al.*, 1958), human (Lee *et al.*, 1961), and porcine ACTH (Shepherd *et al.*, 1956). Glycosylation is followed by cleavage between ACTH and β-LPH, resulting in the formation of glycosylated ACTH intermediates and β-LPH as in AtT-20-D16v cells. A second proteolytic cleavage then occurs to release glycosylated and unglycosylated forms of ACTH (Roberts *et al.*, 1978; Eipper and Mains, 1978; Mains and Eipper, 1979; Rosa *et al.*, 1980; Hinman and Herbert, 1980) and an N-terminal fragment. In the anterior pituitary of mouse and rat, processing essentially ceases at this point, but in the neurointermediate lobe ACTH is converted to α-MSH and CLIP by proteolytic cleavage in the middle of the molecule. The N-terminal portion of ACTH is then trimmed back to 13 residues presumeably by carboxypeptidases, amidated at the C-terminus, and acetylated at the N-terminus (Scott *et al.* 1973, 1974; Jackson and Lowry, 1980). The β-LPH portion of POMC is cleaved to form β-LPH and β-endorphin. β-Endorphin is then acetylated at its N-terminus and shortened by removal of four or five C-terminal amino acids (Zakarian and Smyth, 1980; Smyth *et al.*, 1979; Eipper and Mains, 1981).

While β-endorphin is present in both lobes of the pituitary and the brain, the form of this peptide differs in these tissues (Smyth and Zakarian, 1980; Liotta *et al.*, 1981; Eipper and Mains, 1981; Weber *et al.*, 1982). In the anterior pituitary, almost all of the β-endorphin is β-endorphin (1–31). In the intermediate lobe, this form of endorphin represents a minor component with the short forms [endorphin (1–26) and endorphin (1–27)] predominating. A large fraction of all forms of β-endorphin in the intermediate pituitary is N-acetylated. This is an important modification because it eliminates the analgesic activity of endorphins (Smyth *et al.*, 1979; Deakin *et al.*, 1980).

Differences in processing of POMC peptides have also been observed in various parts of the brain. In the hypothalamus, β-endorphin (1–31) is the major form of endorphin present, whereas *N*-acetyl-endorphin [(1–27) and (1–26)] account for most of the endorphin in the hippocampus and brain stem (Zakarian and Smyth, 1982). However, N-acetylated forms of endorphin account for only a small fraction of total brain endorphin (Weber *et al.*, 1982). ACTH (1–39) and various form of ACTH (1–13) are also reported to have a differential distribution in the brain (DeWied and Jolles, 1982). Modifications of ACTH (1–13) are very important because they greatly influence the bioactivity of the peptide.

Extensive processing of the N-terminal portion of POMC also occurs in the pituitary to yield a variety of fragments containing the γ-MSH sequence and a joining peptide between ACTH and the N-terminal region. This subject has been reviewed recently (Civelli *et al.*, 1984) and will not be treated further here.

It is apparent from the above studies that almost all of the peptides derived from POMC arise by proteolytic cleavage at Lys-Arg sites (Fig. 1). It appears that in mammals this sequence is preferred as a cleavage site over Arg-Arg or Lys-Lys.

VII. Approaches to the Identification of Prohormone Processing Enzymes

An important question that arises is what is the nature of the specificity of the enzymes that cleave the basic pairs of amino acids in the precursors and modify the resulting peptides. Attempts to isolate these enzymes and to study their specificity have resulted in the identification of several candidates for neuropeptide processing enzymes. The best-characterized candidate processing enzymes are those that are easy to assay for. A carboxypeptidase B-like enzyme, for example, involved in enkephalin processing has been purified and characterized because of an extremely simple assay that has been developed (Fricker, 1985). Acetylating and amidating enzymes have also been characterized because of the ease of assay (Bradbury *et al.*, 1982; Eipper *et al.*, 1983; Glembotski, 1982; Mains and Eipper, 1984). The trypsin-like endoproteases that make the initial cleavages at pairs of basic amino acid residues have been much more difficult to study because of the difficulty of the assay and the presence of many kinds of endoproteases in the cell, particularly in the lysosomal fraction. Attempts to overcome these problems by isolating secretory granules have not been completely successful because of contamination of these granules with lysosomes that have very high levels of peptidases. Thus, the biochemical approach has had limited success.

A. Gene Transfer Systems

Experiments with isolated enzymes and substrates are very important in defining the range of specificity of a protease. However, if one wishes

3. BIOSYNTHESIS OF ACTH

to know the specificity of an enzyme in the environment in which it operates in the cell, experiments with isolated enzymes are of limited value. Recently, gene transfer approaches have been developed that allow one to introduce prohormone genes into different kinds of cells and evaluate the capacity of these cells to process the prohormone. Genes encoding growth hormone (Pavlakis et al., 1981) and preproinsulin (Gruss and Khoury, 1981; Lomedico et al., 1984) have been expressed in monkey kidney cells which do not have a regulated secretory pathway consisting of Golgi and secretory vesicles. These cells integrate the transfected genes into their genomes and produce mature mRNA and precursor protein. However, they do not process the protein beyond removal of the signal sequence. Likewise, removal of the signal sequence is the only processing event detected when cDNA encoding proinsulin or growth hormone is expressed in monkey kidney cells or fibroblasts (Robins et al., 1982; Laub and Rutter, 1983). However, when a pituitary cell line, such as the AtT-20 or the GH_4 line, is used as the recipient system for transfection of cDNA, proinsulin and proparathyroid hormone are processed to mature peptides (Moore et al., 1983; Hellerman et al., 1984). Hence, cells without regulated secretory pathways are restricted to removal of the signal peptide, whereas some neurosecretory cells possess ability to process different precursors.

We have used the gene transfer approach to determine if a neurosecretory cell that secretes neuropeptides can process and secrete a foreign neuropeptide precursor in a manner analogous to the endogenous precursor. We have chosen the AtT-20 cell as the recipient cell system for these experiments because this cell produces large quantities of POMC. The steps involved in the processing of POMC to ACTH, β-LPH, β-endorphin, and MSH are well characterized in these cells, and the regulation of secretion of these peptides by CRF, glucocortioids, and cAMP compounds is defined. Thus, by gene transfer, we have presented these cells with the gene that encodes human proenkephalin, a protein that shares many structural features with POMC, including a large number of potential endoproteolytic cleavage sites (12 in all) (Comb et al., 1985).

A plasmid was constructed by subcloning the human proenkephalin gene into PBR 322 (Comb et al., 1985). Cotransformation of AtT-20 cells with the above plasmid and a plasmid containing the selectable marker for neomycin resistance (pRSVneo) led to isolation of 25 clones containing one or more copies of the proenkephalin gene integrated into the host genome. Northern blot analyses showed that several of those clones expressed high levels of a 1.4 kb mRNA identical in size to mature human proenkephalin mRNA. To determine if proenkephalin protein is produced by the transformed clones, peptides were extracted, digested with trypsin and carboxypeptidase, and assayed for Met-enkephalin immunoreactivity (IR). Some clones were found to have very high levels of Met-enkephalin IR, in some cases matching or exceeding levels of the endogenous peptide ACTH (Comb et al., 1985).

The next step was to determine how extensively proenkephalin is processed in the transformed cells. Gel exclusion chromatography of the extracted peptides prior to digestion with proteolytic enzymes indicated that most of the Met-enkephalin IR material was present as low-molecular-weight peptides—less than 3,000 (Comb et al., 1985). Further analysis of the low-molecular weight fraction by reverse-phase HPLC showed that almost all of this material (80–90%) migrated with authentic Met-enkephalin.

Hence, AtT-20 cells are capable of cleaving almost all of the pairs of basic amino acid residues in human proenkephalin (Fig. 6). This is of considerable interest because these cells cleave only four of the eight pairs of basic amino acid residues in mouse POMC to produce ACTH, β-LPH, an N-terminal fragment, and a joining fragment (Douglass et al., 1984) (Fig. 6).

Because the human proenkephalin introduced into these cells appears to be transcribed, translated, and processed in a conventional manner, we have asked whether secretion of Met-enkephalin is similar to that of the endogenous peptides derived from POMC. Table I shows the amounts of ACTH and Met-enkephalin IR released under basal conditions and after treatment of cells with corticotropin releasing factor (CRF) or dexamethasone for 45 minutes. CRF stimulated secretion of both ACTH and Met-enkephalin IR approximately 2-fold above the basal release level. Short-

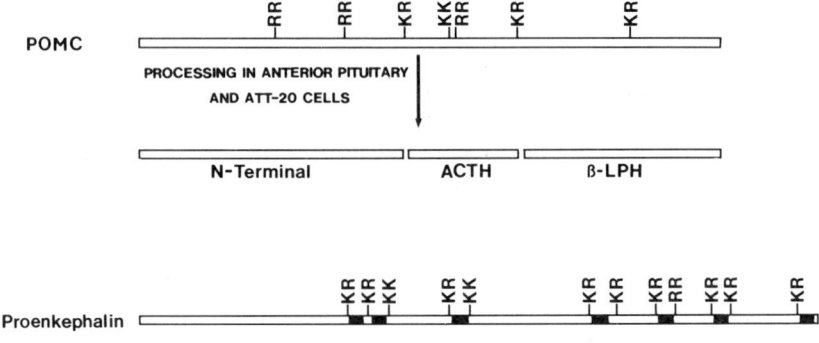

FIG. 6. Schematic comparision of the primary protein structure of mouse POMC and human proenkephalin. Only two or three of the eight pairs of basic amino acids appearing within the sequence of POMC are cleaved by AtT-20 cells at a significant rate, whereas all of these sites are cleaved in proenkephalin.

Table I—Secretion of Met-Enkephalin IR and ACTH IR by AtT-20/hENK Clone d

	pmol IR/ml media[a]	
	ACTH	Met-enkephalin
Control	2.56	0.98
CRF (10^{-8} M)	5.40	2.61
DEX (10^{-6} M)	2.80	1.15

[a] Cells were grown in 24-well plates (Costar) as described in the text. Values represent the average of three wells, each assayed in duplication. Incubation was carried out for 45 min in the presence or absence of regulators.

term treatment with dexamethasone did not induce or inhibit secretion of either peptide, in agreement with results reported by others (Yates and Maran, 1974). The molar ratio of the secreted peptides remained relatively constant with varying degrees of release, suggesting that the two peptides may be contained within the same population of secretory vesicles.

Finally, we wanted to know how levels of human proenkephalin mRNA are regulated in the transformed cells. Previous studies have shown that the level of POMC mRNA in AtT-20 cells is elevated by treatment of these cells with CRF or cyclic AMP and depressed by long-term treatment with dexamethasone. Proenkephalin is normally expressed in chromaffin cells from bovine adrenal medulla as well as in a number of sites in the brain and reproductive tract. In bovine chromaffin cells, it has been shown that proenkephalin mRNA is elevated by agents that increase cyclic AMP levels, such as forskolin and by nicotinic cholinergic agonists (Eiden et al., 1984). We have found that cyclic AMP and CRF increase the levels of human proenkephalin mRNA severalfold in transformed AtT-20 cells (unpublished studies). Hence, the transfected proenkephalin gene contains the sequences that are required for regulation of expression of this gene by cyclic AMP and CRF.

B. Use of Vaccinia Virus as a Transformation Vehicle

The above method of gene transfer leads to integration of transfected genes into the host genome and creates stable cell lines which are very useful for the study of transcriptional regulation of gene expression. If one wishes to study regulation of expression of a gene product in the cytoplasm of a cell, it is easier to introduce cDNA directly into the cytoplasm of a cell rather than into the host nuclear genome. Cells remain viable for many hours or days after infection. This approach is now possible because of the development of vaccinia virus as a cloning and expression vehicle as mentioned in the previous section. Use of this virus as an expression vector has several major advantages: (1) the virus has a very broad host cell range, and (2) unlike other DNA viruses, the infectious cycle occurs entirely in the cytoplasm of the host cell. Insertion of cDNA into the virus downstream from the early viral promotor leads to rapid transcription of the cDNA in the host cell (within minutes of infection) and efficient production of protein from the mRNA. Although the use of vaccinia virus as an expression vector is relatively new, several viral coat proteins have already been produced by this approach (Smith et al., 1983a,b; Bennink et al., 1984; Panicali et al., 1983; Wiktar et al., 1984).

The construction of the recombinant vaccinia virus containing human proenkephalin cDNA (VV:PE) has already been described as well as the titer of virus required to produce maximum expression of Met-enkephalin in various mammalian cell lines (Hruby et al., 1983). A notable feature of this construction is the presence of an early viral promotor upstream from the DNA.

A salient feature of the vaccinia virus expression vector system is the ability to infect a wide spectrum of cell types and observe the production of a foreign protein. To this end, five different cell lines were infected with VV:PE. After 24 hr of infection, the cells were harvested and extracted with acetic acid. Proteins and peptides in the extract were digested sequentially with trypsin and carboxypeptidase B to release enkephalin from larger peptides and then assayed for Met-enkephalin IR by radioimmunoassay. The results, presented in Table II, show that although Met-enkephalin IR was detected in each cell type, the level of Met-enkephalin IR varied over 5-fold in both the cellular and secreted levels depending on the cell type infected.

Although infection of each cell type with VV:PE resulted in the production of Met-enkephalin IR, it was not known to what extent each cell type was capable of processing human proenkephalin into smaller peptides. In order to answer this question, acetic acid extracts from cells and culture medium were lyophilized and resuspended in 0.25 M triethylammonium formate. Each sample was then applied to a TSK-125 HPLC sizing column. Following peptide separation, each fraction was assayed for Met-enkephalin IR. The results are presented in Fig. 7 for AtT-20 cells and BSC-40 cells.

Extracts of AtT-20 cells infected with VV:PE exhibit five major peaks

Table II—Met-Enkephalin Immunoradioactivity[a]

	BSC40 (0.5 PFU/cell)[b]	AtT-20D16v (5 PFU/cell)	GH4 (5 PFU/cell)	L (5 PFU/cell)	P388D$_1$ (5 PFU/cell)
MI	—	—	—	—	—
VV:WT	—	0.1[c]	—	—	—
VV:PE					
(Cell)	19.4	10.1	43.6	7	18.3
(Secreted)	180.0	36.0	79.0	43	70.0

[a] Parallel plates of each cell line were either mock infected (MI) or infected with VV:WT or VV:PE at the indicated titer. After 24 hr of infection, cells were processed for the Met-enkephalin radioimmunoassay (RIA).
[b] See Hruby et al. (1983) for definitions of plaque-forming units (PFU).
[c] Values are given as pmol Met-enkephalin in IR/10⁶ cells. Each value is the average of two separate experiments.

of Met-enkephalin IR (Fig. 7, top). Peak 1 Met-enkephalin IR material migrates with an apparent molecular weight of 28,500, which is the size of human proenkephalin. Peak 2 Met-enkephalin IR material migrates with an apparent molecular weight of 16,000. The slowest migrating peak, 5, coelutes with purified Met-enkephalin. Analysis of the media from VV:PE-infected AtT-20 cells reveals two major peaks of secreted Met-enkephalin IR. The faster migrating peak coelutes with peak 1 of the cell extract and the slower migrating peak coelutes with peak 5 and purified Met-enkephalin. Hence, both the plasmid and vaccinia virus method of transfection of the human proenkephalin-coding DNA result in the production of free Met-enkephalin by AtT-20 cells, making the point that infection by vaccinia virus does not alter processing of proenkephalin over the period of these experiments.

In contrast to AtT-20 cells, analysis of BSC-40 cells infected with VV:PE reveals two prominent peaks of Met-enkephalin IR (Fig. 7, bottom) which elute in the same positions as peak 1 in VV:PE-infected AtT-20 cells (Fig. 7, top). Similar results were obtained for GH$_4$C1, LtK, and P388D$_1$ cell lines (data not shown). In order to determine whether Met-enkephalin IR material in peaks 1 and 2 is secreted, a small aliquot of the culture medium from BSC-40 cells was also analyzed. As shown in Fig. 7 (bottom), only one prominent peak of Met-enkephalin IR is secreted. This peak coeluted with peak 1 from the cell extract.

A recombinent vaccinia virus containing POMC cDNA has also been constructed and used to transform the same cell lines listed in Table II. Processing of POMC was analyzed in the cells and the medium, and the

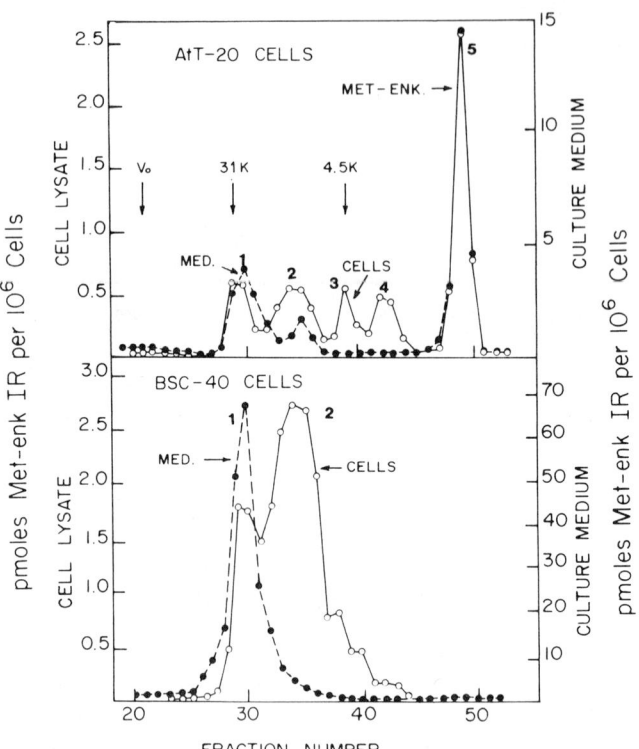

FIG. 7. Processing of human proenkephalin in AtT-20 cells (top) and BSC-40 cells (bottom) transformed by vaccinia virus. A small aliquot of either acetic acid cell extracts (○) or culture medium (●) was dried by rotor evaporation (speed vac, Savant) and resuspended in 100 μl of 0.25 M triethylammonium formate (TEAF), pH 3.0. The resuspended sample was sonicated (VibraCell, Sonics Materials) and insoluble material was removed by centrifugation. Each sample was next applied to a RSK-125 Bio-Sil HPLC sizing column. The column was run in 0.25 M TEAF, pH 3.0, at a flow rate of 0.5 ml/minute. Fraction volumes were 300 μl. Following removal of the TEAF, each fraction was digested with trypsin and carboxypeptidase B. An aliquot of each digest was assayed for Met-enkephalin IR (Met-enk IR). Met-ENK, Met-enkephalin.

same results were obtained as for proenkephalin; that is, the only cell line that showed extensive processing of POMC to ACTH-size material was the AtT-20 cell line.

VIII. Conclusions

The discovery of POMC was the key to elucidation of the biosynthetic pathway of ACTH, melanocyte stimulating hormones, β-LPH, and β-endorphins. POMC was the first member of a new class of proteins that contain the sequences of more than one bioactive peptide. These proteins, referred to as "polyproteins," are expressed in a wide variety of tissues where they undergo differential processing (Douglass et al., 1984). POMC, for example, is processed mainly to ACTH and β-LPH in the anterior pituitary, whereas in the neurointermediate pituitary it is processed to α-MSH. β-MSH, β-endorphin, and acetylated and C-terminally shortened forms of β-endorphin. POMC is also formed in several regions of the brain where it also undergoes differential processing.

The regulation of production of POMC is also different in the two lobes of the pituitary. Glucocorticoids and CRF regulate expression of POMC peptides at the level of both gene transcription and secretion (Birnberg et al., 1983), whereas in the neurointermediate lobe these processes are regulated by catecholamines and not by steroids and CRF.

Thus, the regulation of POMC expression illustrates in a dramatic way the potential that exists in a eukaryotic cell for creating diversity in the products generated from the expression of a single gene.

The sequencing of the POMC gene and cDNA from several species by recombinant DNA techniques indicates that the arrangement of exons and introns in this gene is highly conserved. Likewise, the arrangement of all of the bioactive domains in the protein is highly conserved in all these species (which include five mammals, an amphibian, and a fish). Amino acid sequence homology is very high for the MSH and β-endorphin domains in POMC. Pairs of basic amino acids delineate the bioactive domains in all species, suggesting that the cleavage enzymes in different organisms have similar specificity. However, a number of recent results indicates that not all of these cleavage sites are equivalent. For example, only four of eight pairs of basic amino acid residues in POMC are cleaved in AtT-20 cells and anterior pituitary cells, whereas all of these sites are cleaved in the neurointermediate lobe of the pituitary. It is not clear from this whether selectivity for cleavage sites is due to differences in cleavage enzymes or differences in the intracellular environment that may alter enzyme specificity or substrate structure or both.

Thus far it has been very difficult to characterize endoproteolytic processing enzymes and to study their specificity. However, some new approaches have been developed recently that will help define this specificity. The genes or cDNA encoding neuroendocrine precursors can be introduced into a wide variety of cell types by gene transfer techniques. This allows one to compare the ability of different cells (secretory and nonsecretory) to process a given precursor. In this chapter we have shown how we can introduce the human proenkephalin gene into mouse AtT-20 cells which normally produce POMC but not proenkephalin. The human proenkephalin gene is expressed at a high level in these cells and is processed almost completely to free Met-enkephalin. Hence, in the same cellular environment in which only 4 of 8 potential cleavage sites in POMC are cut, all 10 such cleavage sites in human proenkephalin are used. This finding shows that features of structure other than pairs of basic amino acids are important in dictating which sites are cleaved.

We have also used vaccinia virus as a vector for introducing human proenkephalin into cells to study processing. This vector has the advantage over plasmids in that it is able to transform a wide variety of cell types with very high efficiency. With this approach, we have been able to show that synthesis of the same precursor, proenkephalin, can occur in many types of cells but processing to Met-enkephalin is very specific and occurs in only one of the five cell types tested (Table II). Another important point is that, although BH_4Cl cells do not process proenkephalin to Met-enkephalin, they are capable of converting other precursor proteins to mature bioactive peptides. For example, GH_4Cl cells that have been transformed with proparathyroid hormone cDNA can convert this precursor to mature parathyroid hormone. $P388D_1$ cells secrete interleukin 1 in the form of an inactive precursor protein (31 kDa precursor)(Lomedico et al., 1984) which is cleaved during or shortly after secretion to form mature interleukin. In both precursors, proteolytic cleavage sites exist N-terminal to the sequence of the mature peptide. The cleavage sequence is Lys-Lys-Arg in proparathyroid hormone and Lys-Lys-Arg-Arg in the interleukin 1 precursor. If these sequences are actually the sites of cleavage in these precursors, then the enzyme systems involved must require more than pairs of basic amino acid residues as cleavage recognition signals, otherwise they would process proenkephalin to enkephalin peptides. This suggests that endoproteolytic processing enzymes have a high degree of specificity and that a few enzymes with broad specificity might not be enough to process the wide variety of neuroendocrine precursors that exist.

Future studies with gene transfer systems may help to further resolve questions about specificity of processing enzymes. For example, it is possible to reduce the level of a specific protein in a cell by the introduction of antisense cDNA into a cell by transfection (Izant and Weintraub, 1985). The antisense cDNA produces antisense mRNA that interferes with translation of the sense mRNA, thus reducing the level of the protein after a period of turnover. If one has available cDNA clones encoding potential processing enzymes, such as exo- or endoproteases, one can introduce the antisense strand of the cDNA encoding these enzymes into a cell that expresses a neuropeptide and determine what effect the reduction in level of the enzyme has on specific steps involved in processing the precursor.

Another possibility is to use a vector like vaccinia virus to introduce cDNA coding for candidate processing enzymes and neuropeptide precursors into different types of null cells (nonprocessing cells as shown in Table II) and determine the effect expression of the enzyme has on processing of the precursor. The broad host range of vaccinia virus makes it a good candidate for the transformation vector in these experiments.

One can also use gene transfer methods to produce large quantities of a neuropeptide precursor as well as large quantities of processing enzymes. With purified enzymes and natural substrates, one could carry out detailed studies of specificity of enzymes. One could then make amino acid substitutions in the substrate and the enzyme by *in vitro* mutagenesis techniques and determine the effect of these alterations on specificity of cleavage reactions in the cell.

The combined use of the approaches outlined above should provide a very powerful arsenal of techniques for probing the specificity of processing enzymes and the factors that regulate neuropeptide production in the next 5 to 10 years.

Acknowledgments

We thank L. Williams and N. Gay for expert manuscript preparation. G. T. was supported by a Damon Runyon–Walter Winchell Cancer Fellowship, DRG-797. D. L. was supported by Postdoctoral Fellowship F32-DA05261. L. F. was supported by PHS Fellowship NS07361-02. M. C., M. M. and N. B. were supported by the University of Oregon Molecular Biology Training Grant. The research performed in the E. H. lab was supported by Research Grants AM-16879 and AM-30155 from the National Institutes of Arthritis, Diabetes, Digestive and Kidney Diseases, and Research Grant DA-02736 from the National Institute on Drug Abuse.

References

Akil, H., Veda, Y., Lin, H. L., and Watson, S. J. (1981). *Neuropeptides* **1**, 429–446.
Baizman, E. R., and Cox, B. M. (1978). *Life Sci.* **22**, 519–524.
Baizman, E. R., Cox, B. M., Osman, O. H., and Goldstein, A. (1979). *Neuroendocrinology* **28**, 402–409.
Bennett, H. P. J., Browne, C. A., and Solomon, S. (1981a). *Biochemistry* **20**, 4530–4538.
Bennett, H. P. J., Browne, C. A., and Solomon, S. (1981b). *Proc. Natl. Acad. Sci. U.S.A.* **78**, 4713–4717.
Bennink, J. R., Yewdell, J. W., Smith, G. L., Moller, C., and Moss, B. (1984). *Nature (London)* **311**, 578–579.
Birnberg, N., Lissitzky, J.-C., Hinman, M., and Herbert, E. (1983). *Proc. Natl. Acad. Sci. U.S.A.* **80**, 6982–6986.
Blobel, G., and Dobberstein, B. (1975a). *J. Cell Biol.* **67**, 835–851.
Blobel, G., and Dobberstein, B. (1975b). *J. Cell Biol.* **67**, 852–862.

Boileau, G., Barbeau, C., Jeannotte, L., Chretien, M., and Drouin, J. (1983). *Nucleic Acids Res.* **11,** 8063–8071.
Bradbury, A. F., Finnie, M. D. A., and Smyth, D. G. (1982). *Nature (London)* **298,** 686–689.
Brahic, M., and Haase, A. T. (1978). *Proc. Natl. Acad. Sci. U.S.A.* **75,** 6125–6129.
Burgers, A. C. J., Imai, K., and vanOordt, G. V. (1983). *Gen. Comp. Endocrinol.* **3,** 53–56.
Chang, A. C. Y., Cochet, M., and Cohen, S. (1980). *Proc. Natl. Acad. Sci. U.S.A.* **77,** 4890–4894.
Chen, C. L. C., Dionne, F. T., and Roberts, J. L. (1983). *Proc. Natl. Acad. Sci. U.S.A.* **80,** 2211–2215.
Civelli, O., Birnberg, N., and Herbert, E. (1982). *J. Biol. Chem.* **257,** 6783–6787.
Civelli, O., Douglass, J., and Herbert, E. (1984). *Peptides* **6,** 70–90.
Civelli, O., Douglass, J., Goldstein, A., and Herbert, E. (1985). *Proc. Natl. Acad. Sci. U.S.A.* **82,** 4291–4295.
Comb, M., Seeburg, P. H., Adelman, J., Eiden, L., and Herbert, E. (1982). *Nature (London)* **295,** 663–666.
Comb, M., Liston, D., Rosen, H., and Herbert, E. (1985). *EMBO J.* **12,** 3115–3122.
Deakin, J. F. W., Dostrousky, O., and Smyth, D. G. (1980). *Biochem. J.* **189,** 501–506.
DeWied, D., and Jolles, J. (1982). *Physiol. Rev.* **62,** 976–1059.
Dorner, A. J., and Kemper, B. (1978). *Biochemistry* **17,** 5550–5555.
Douglass, J., Civelli, O., and Herbert, E. (1984). *Annu. Rev. Biochem.* **53,** 665–715.
Drouin, J., and Goodman, H. M. (1980). *Nature (London)* **288,** 610–613.
Eiden, L. E., Giraud, P., Dave, J. R., Hotchkiss, A. J., and Affolter, H.-U. (1984). *Nature (London)* **312,** 661–663.
Eipper, B. A., and Mains, R. E. (1978). *J. Supramol. Struct.* **8,** 247–256.
Eipper, B. A., and Mains, R. E. (1980). *Endocr. Rev.* **1,** 1–27.
Eipper, B. A., and Mains, R. E. (1981). *J. Biol. Chem.* **256,** 5689–5695.
Eipper, B. A., and Mains, R. E. (1982). *J. Biol. Chem.* **257,** 4907–4915.
Eipper, B. A., Glembotski, C. C., and Mains, R. E. (1983). *J. Biol. Chem.* **258,** 7292–7298.
Evans, R. M., Birnberg, N. C., and Rosenfeld, M. G. (1982). *Proc. Natl. Acad. Sci. U.S.A.* **79,** 7659–7663.
Farah, J. M., Malcolm, D. S., and Mueller, G. P. (1982). *Endocrinology* **110,** 657–659.
Fleischer, N., and Rawls, W. E. (1970). *Am. J. Physiol.* **4,** 367–404.
Fricker, L. D. (1985). *Trends Neurosci.* **8,** 210–214.
Gee, C. E., and Roberts, J. L. (1982). *DNA* **2,** 157–163.
Gianoulakis, C., Seidah, N. G., Routhier, R., and Chretien, M. (1979). *J. Biol. Chem.* **254,** 11903–11906.
Glembotski, C. G. (1982). *J. Biol. Chem.* **257,** 10501–10509.
Gruss, P., and Khoury, G. (1981). *Proc. Natl. Acad. Sci. U.S.A.* **78,** 133–137.
Gubler, U., Seeburg, P., Hoffman, B. J., Gage, L. P., and Udenfriend, S. (1982). *Nature (London)* **295,** 206–208.
Gudelsky, G. A., Nansel, D. D., and Porter, J. C. (1980). *Endocrinology* **107,** 30–34.
Harpold, M. M., Dobner, P. R., Evans, R. M. Bancroft, F. C. and Darnell, J. E. (1978). *Nucleic Acids Res.* **5,** 2039–2053.
Hellerman, J. G., Cone, R. C., Potts, J. T., Rich, A., Mulligan, R. C., and Kronenberg, H. M. (1984). *Proc. Natl. Acad. Sci. U.S.A.* **81,** 5340–5344.
Herbert, E., Allen, R., and Paquette, T. (1978). *Endocrinology* **102,** 218–227.
Herbert, E., Birnberg, N., Lissitzsky, J.-C., Civelli, O., and Uhler, M. (1981). *Neurosci. Newsl.* **1,** 15–27.
Hinman, M., and Herbert, E. (1980). *Biochemistry* **19,** 5395–5402.
Hollt, V., and Bergmann, M. (1982). *Neuropharmacology* **21,** 147–154.

Hollt, V., Haarmann, I., Seizinger, B. R., and Hertz, A. (1982). *Endocrinology* **110**, 1885–1891.
Hook, V. Y. H., Eiden, L. E., and Brownstein, M. J. (1982). *Nature (London)* **295**, 341–342.
Howard, K. S., Shephard, R. G., Eiquer, E. A., Davis, D. S., and Bell, P. H. (1955). *J. Am. Chem. Soc.* **77**, 3449–3420.
Hruby, D. E., Thomas, G., Herbert, E., and Franke, C. A. (1983). In "Methods in Enzymology" (P. M. Conn, ed.), Vol. 103. Academic Press, New York.
Hubbard, S. C., and Ivatt, R. J. (1981). *Annu. Rev. Biochem.* **50**, 555–584.
Hudson, P., Penschow, J., Shine, J., Ryan, G., Niall, H., and Coghlan, J. (1981). *Endocrinology* **108**, 353–356.
Izant, J. C., and Weintraub, H. (1985). *Science* **229**, 346–352.
Jackson, R. C., and Blobel, G. (1977). *Proc. Natl. Acad. Sci. U.S.A.* **74**, 5598–5602.
Jackson, S. and Lowry, P. J. (1980). *Ann. N.Y. Acad. Sci.* **86**, 205–219.
Kakidani, H., Furutani, Y., Takahashi, H., Noda, M., Morimoto, Y., Hirose, T., Asai, M., Inayama, S., Nakanishi, S., and Numa, S. (1982). *Nature (London)* **298**, 245–249.
Kraicer, J. (1977). *Front. Horm. Res.* **4**, 200–222.
Krieger, D. T., Liotta, A., and Brownstein, M. J. (1977). *Proc. Natl. Acad. Sci. U.S.A.* **74**, 648–652.
Krieger, D. T., Liotta, A. S., Brownstein, M. J., and Zimmerman, E. A. (1980). *Recent Prog. Morm. Res.* **36**, 277–344.
Langer-Safer, P. R., Levine, M., and Ward, D. C. (1982). *Proc. Natl. Acad. Sci. U.S.A.* **79**, 4381–4385.
Larsson, L. I. (1977). *Lancet* **2**, 1321–1323.
Laub, O., and Rutter, W. J. (1983). *J. Biol. Chem.* **258**, 6043–6050.
Lee, T. H., Lerner, A. B., and Buettner-Janusch, V. (1961). *J. Biol. Chem.* **236**, 2970–2974.
Lepine, J., and Dupont, A. (1981). *Endocr. Soc.* **108**, 385 (Abstr.).
Li, C. H., Geschwind, I. I., Levy, A. L., Harris, J. I., Dixon, J. S., Pon, N. G., and Porath, J. O. (1954). *Nature (London)* **173**, 251–253.
Li, C. H., Geschwind, I. I., Cole, R. D., Raacke, I. D., Harris, J. I., and Dixon, J. S. (1955). *Nature (London)* **176**, 687–689.
Li, C. H., Dixon, J. S., and Chung, D. (1958). *J. Am. Chem. Soc.* **80**, 2587–2588.
Lingappa, V. R., and Blobel, G. (1980). *Recent. Prog. Horm. Res.* **36**, 451–475.
Liotta, A. S., Yamaguchi, H., and Krieger, D. T. (1981). *J. Neurosci.* **1**, 585–595.
Lissitzky, J.-C., Morin, O., Dupont, A., Labrie, F., Seidah, N. G., Chretien, M., Lis, M., and Coy, D. H. (1978). *Life Sci.* **22**, 1715–1726.
Loeber, J. G., Verhoef, J., Burbach, J. P. H., and Wilter, A. (1979). *Biochem. Biophys. Res. Commun.* **86**, 1288–1295.
Lomedico, P. T., Gubler, U., Hellmann, C. P., Dukovich, M., Giri, J. G., Pan, Y.-C. E., Collier, K., Semionow, R., Chua, A. O., and Mizel, S. B. (1984). *Nature (London)* **312**, 458–462.
Mains, R. E., and Eipper, B. A. (1979). *J. Biol. Chem.* **254**, 7885–7894.
Mains, R. E., and Eipper, B. A. (1984). *Endocrinology* **115**, 1683–1690.
Mains, R. E., Eipper, B. A., and Ling, N. (1977). *Proc. Natl. Acad. Sci. U.S.A.* **74**, 3014–3018.
Marshall, R. D. (1974). *Biochem. Soc. Symp.* **40**, 17–26.
Martens, G., Civelli, O., and Herbert, E. (1985). *J. Biol. Chem.* **260**, 13685–13689.
Moon, H. D., Li, C. H., and Jennings, B. M. (1973). *Anat. Rec.* **175**, 529–538.
Moore, H. P. H., Walker, M. D., Lee, F., and Kelly, R. B. (1983). *Cell* **35**, 531–538.
Nakai, Y., Nakao, K., Oki, S., and Imura, H. (1978). *Life Sci.* **23**, 2013–2018.

Nakamura, M., Nakanishi, S., Sueoka, S., Imura, H., and Numa, S. (1978). *Eur. J. Biochem.* **86,** 61–66.
Nakanishi, S., Kita, T., Taii, S., Imura, H., and Numa, S. (1977). *Proc. Natl. Acad. Sci. U.S.A.* **74,** 3283–3286.
Nakanishi, S., Inoue, A., Kita, T., Nakamura, M., Chang, A. C. Y., Cohen, S. N., and Numa, S. (1979). *Nature (London)* **278,** 423–427.
Nakanishi, S., Teranishi, Y., Watanabe, Y., Notake, M., Noda, M., Kakidani, H., Jingami, H., and Numa, S. (1981). *Eur. J. Biochem.* **115,** 429–438.
Nansel, D. D., Gudelsky, G. A., and Porter, J. C. (1979). *Endocrinology* **105,** 1073–1077.
Noda, M., Furutani, Y., Takahashi, H., Toyosato, M., Hirose, T., Inayama, S., Nakanishi, S., and Numa, S. (1982). *Nature (London)* **295,** 202–206.
Notake, M., Tobimatsu, T., Watanabe, Y., Takahashi, H., Mishina, M., and Numa, S. (1983). *FEBS Lett.* **156,** 67–71.
Oates, E., and Herbert, E. (1984). *J. Biol. Chem.* **259,** 7421–7425.
Orwoll, E. S., and Kendall, J. W. (1980). *Endocrinology* **107,** 438–442.
Panicali, D., Davis, S. W., Weinberg, R. L., and Paoletti, E. (1983). *Proc. Natl. Acad. Sci. U.S.A.* **80,** 5364–5368.
Patzelt, C., Labrecque, A. D., Duguid, J. R., Carroll, R. J., Keim, P., Heinrikson, R. L., and Steiner, D. F. (1978). *Proc. Natl. Acad. Sci. U.S.A.* **75,** 1260–1264.
Pavlakis, G. N., Hizuka, N., Gorden, P., Seeburg, P., and Hamer, D. H. (1981). *Proc. Natl. Acad. Sci. U.S.A.* **78,** 7398–7402.
Pedersen, R. C., and Brownie, A. C. (1980). *Proc. Natl. Acad. Sci. U.S.A.* **77,** 2239–2243.
Pelletier, G., Lederc, R., Labrie, F., Cote, J., Chretien, M., and Lis, M. (1977). *Endocrinology* **100,** 770–776.
Policastro, P., Phillips, M., Oates, E., Herbert, E., Roberts, J. L., Seidah, N., and Chretien, M. (1981). *Eur. J. Biochem.* **116,** 255–259.
Raese, J. D., Boarder, M. R., Makk, G., and Barchas, J. D. (1980). *Adv. Biochem. Psychopharmacol.* **22,** 377–383.
Rees, H., Stumpf, W., Sar, M., and Petrusz, P. (1977). *Cell Tissue Res.* **182,** 347–356.
Roberts, J. L., and Herbert, E. (1977a). *Proc. Natl. Acad. Sci. U.S.A.* **74,** 4826–4830.
Roberts, J. L., and Herbert, E. (1977b). *Proc. Natl. Acad. Sci. U.S.A.* **74,** 5300–5304.
Roberts, J. L., Phillips, M., Rosa, P. A., and Herbert, E. (1978). *Biochemistry* **17,** 3609–3618.
Roberts, J. L., Budarf, M. L., Baxter, J. D., and Herbert, E. (1979). *Biochemistry* **18,** 4907–4915.
Roberts, J. L., Chen, C. L. C., Dionne, F. T., and Gee, C. E. (1982). *Trends Neurosci.* **11,** 314–317.
Robins, D. M., Pack, I., Seeburg, P. H., and Axel, R. (1982). *Cell* **29,** 623–631.
Rosa, P., Policastro, P., and Herbert, E. (1980). *J. Exp. Biol.* **89,** 215–237.
Schachter, B. S., Johnson, L. K., Baxter, J. D., and Roberts, J. L. (1983). *Endocrinology* **110,** 1442–1444.
Scott, A. P., Ratcliff, J. G., Rees, L. H., Bennett, H. P. J., Lowry, P. J., and McMartin, C. (1973). *Nature (London) New Biol.* **244,** 65–67.
Scott, A. P., Lowry, P. J., Ratcliff, J. G., Rees, L. H., and Landon, J. (1974). *J. Endocrinol.* **61,** 355–364.
Scott, A. P., Lowry, P. J., van Wimersma, and Greidanus, T. B. (1976). *J. Endocrinol.* **70,** 197–205.
Seidah, N. G., Rochemont, J., Hamelin, J., Lis, M., and Chretien, M. (1981a). *J. Biol. Chem.* **256,** 7977–7984.

Seidah, N. G., Rochemont, J., Hamelin, J., Benjannet, S., and Chretien, M. (1981b). *Biochem. Biophys. Res. Commun.* **102,** 710–716.
Seizinger, B. R., and Hollt, V. (1980). *Biochem. Biophys. Res. Commun.* **96,** 535–543.
Shepherd, R. G., Willson, S. D., Howard, K. S., Bell, P. H., Davis, S. B., Eigner E. A., and Shakespeare, N. E. (1956). *J. Am. Chem. Soc.* **78,** 5067–5076.
Shields, B., and Blobel, G. (1978). *J. Biol. Chem.* **253,** 3753–3756.
Smith, G. L., Mackett, M., and Moss, B. (1983a). *Nature (London)* **302,** 490–495.
Smith, G. L., Murphy, B. R., and Moss, B. (1983b). *Proc. Natl. Acad. Sci. U.S.A.* **80,** 7155–7159.
Smyth, D. G., and Zakarian, S. (1980). *Nature (London)* **288,** 613–615.
Smyth, D. G., Massey, D. E., Zakarian, S., and Finnie, M. D. A. (1979). *Nature (London)* **279,** 251–254.
Soma, G.-I., Kitahara, N., Nishizawa, T., Nanami, H., Kotaki, C., Okazaki, H., and Andoh, T. (1984). *Nucleic Acids Res.* **12,** 8029–8041.
Takahashi, H., Hakamata, Y., Watanabe, Y., Kikuno, R., Miyata, T., and Numa, S. (1983). *Nucleic Acids Res.* **11,** 6847–6858.
Tsong, S. D., Phillips, D., Halmi, N., Liotta, A. S., Margioris, A., Bardin, C. W., and Krieger, D. T. (1982). *Endocrinology* **110,** 2204–2206.
Uhler, M., and Herbert, E. (1983). *J. Biol. Chem.* **258,** 257–261.
Uhler, M., Herbert, E., D'Eustachio, P., and Ruddle, F. (1983). *J. Biol. Chem.* **258,** 9444–9453.
Vale, W., Speiss, J., Rivier, C., and Rivier, J. (1981). *Science* **213,** 1394–1397.
Vermes, I., Mulder, G. H., Smelik, P. G., and Tilders, F. J. H. (1980). *Life Sci.* **27,** 1761–1768.
Warembourg, M. (1975). *Cell Tissue Res.* **161,** 183–191.
Watanabe, H., Nicholson, W. E., and Orth, D. N. (1973). *Endocrinology* **93,** 411–416.
Weber, E., Evans, C. J., Chang, J.-K., and Barchas, J. D. (1982). *J. Neurochem.* **38,** 436–447.
Wiktar, T. J., Macfarlan, R. I., Reagan, K. J., Dietzschold, B., Curtis, P. J., Wunner, W. H., Kieny, M.-P., Lathe, R., Lecocq, J.-P., Mackett, M., Moss, B., and Koprowski, H. (1984). *Proc. Natl. Acad. Sci. U.S.A.* **81,** 7194–7198.
Yates, F. E., and Maran, J. W. (1974). *Handb. Physiol.* **4,** 367–404.
Zakarian, S., and Smyth, D. G. (1980). *Nature (London)* **288,** 613–615.
Zakarian, S., and Smyth, D. G. (1982). *Nature (London)* **296,** 250–253.

4

ACTH and Corticosteroidogenesis

PETER F. HALL

Worcester Foundation for Experimental Biology
Shrewsbury, Massachusetts 01545

ACTH stimulates steroid synthesis in adrenal cells by way of cyclic AMP, the production of which it stimulates by binding to its receptor in the plasma membrane. One of the major unsolved problems of adrenal steroid synthesis lies in the question of whether or not Ca^{2+} serves as a second messenger for ACTH. Cyclic AMP phosphorylates a number of adrenal proteins including cytoplasmic proteins, plasma membrane proteins, and cytoskeletal proteins. The only adrenal protein whose function is known to change as a result of phosphorylation is cholesterol ester hydrolase, which becomes more active when phosphorylated and releases cholesterol from depots of cholesterol ester. Changes in the cytoskeleton, probably arising from phosphorylation of proteins, reorganize the structure of the cell in such a way that the transport of cholesterol to the outer mitochondrial membrane is increased. It is likely that cholesterol travels in the company of a carrier proteins (SCP_2). From this point on newly synthesized protein is required to secure the loading of the C_{27} side-chain cleavage enzyme in the inner mitochondrial membrane. How this is brought about provides another unsolved problem, but the cholesterol may continue its journey bound to SCP_2. Provided with substrate the enzyme now catalyzes a burst of side-chain cleavage resulting in rapid production of pregnenolone which is followed by a much slower rate of production. The reason for this change of pace provides another mystery. Many other provocative findings have been made concerning the response to cyclic AMP and hence to ACTH. These findings are discussed here, but they cannot be fitted into the jigsaw puzzle at this time. This does not mean that such pieces are less important than those already in place; it simply means that we must extend the present scene to a state that will accommodate them.

I. Introduction

The remarkable properties of the adrenal cortex and its secretory products, the corticosteroids, should not cause us to forget that the cells of the cortex face problems common to all cells. The adrenal cortex must be capable of a certain basal level of secretory activity because the glucocorticoids are essential to life, in large measure because they "permit" an enormous array of metabolic activities elsewhere in the body—activities that cannot take place without these hormones (Bush, 1962). In addition, the cortex must take its place in the circadian rhythms of the body, and above all else it must be capable of extremely rapid responses to those unexpected challenges to homeostasis which we refer to collectively as stress (Selye, 1954). The cortex must also be able to return to the resting level of activity when the stressor stimulus is withdrawn. Finally, the adrenal cortex must be capable of prolonged responses to chronic stress with increase in size (Selye, 1954). So far as we know, the only contribution of the adrenal cortex to the economy of the whole organism is the regulated production of steroid hormones; the process of regulation is expressed by changes in the output of steroids from the gland in response to changes in the requirements for such steroids. This chapter considers the role of ACTH in these processes of regulation and, in particular, with the production of glucocorticoids—the characteristic products of the zona fasciculata.

II. Production of Steroids by the Adrenal Cortex

To synthesize steroids the cell requires a substrate or source of the steroid ring system, a series of enzymes and cofactors necessary to transform the substrate to the desired product (glucocorticoids), and the energy necessary to drive endergonic synthetic reactions.

A. The Substrate

The immediate source of the steroid ring system consists of depots of cholesterol in the cytoplasm of the adrenal cell. In these depots cholesterol exists as esters (Gwyne *et al.*, 1976; Verschoor-Klootwyk *et al.*, 1982; Anderson and Dietschy, 1978). The cholesterol can in turn be synthesized from acetate in the adrenal cell (Vahouny *et al.*, 1985) or it can be taken up ready-made from the plasma in the form of lipoprotein particles in which the cholesterol is largely esterified—especially to linoleate (Goldstein and Brown, 1974; Faust *et al.*, 1977; Hall and Nakamura,

1979). The cholesterol ester taken up in this way is deesterified in lysosomes (Hall and Nakamura, 1979). The free cholesterol is presumably available for immediate use, but most of it is reesterified and stored in lipid droplets (Vahouny et al., 1985).

Much of our understanding of these events comes from analogy with the pathways described in fibroblasts (Goldstein and Brown, 1974), where biochemical approaches have been supported by elegant genetic studies in which mutations have demonstrated the consequences of defects in the pathway described in Goldstein et al. (1976). Less work has been done with the adrenal, but it has been possible to show that the same sequence of events occurs in that tissue (Anderson and Dietschy, 1978; Hall and Nakamura, 1979; Carr and Simpson, 1981; Gwynne and Strauss, 1982).

The adrenal cell is also capable of synthesizing its own cholesterol from acetate, and in this way acetate is incorporated into steroids (Karaboyas and Koritz, 1965). For fibroblasts the interrelationships between these two alternative pathways are clear. When cholesterol is available to the cell in the form of low-density lipoprotein, the synthesis of cholesterol from acetate is inhibited by a negative feedback effect on 2-hydroxy-2-methylglutaryl-CoA reductase (HMG CoA reductase) (Brown and Goldstein, 1976). In the absence of adequate exogenous cholesterol or in the presence of genetic defects in handling exogenous cholesterol, the synthesis of cholesterol from acetate increases (Goldstein and Brown, 1974). Such an interplay between the two sources of cholesterol also occurs in the adrenal (Vahouny et al., 1983).

The question of how the adrenal cell uses these various alternative substrates to make glucocorticoids is complicated by important species differences. The rat adrenal stores large amounts of cholesterol ester which it can acquire by the endocytosis of low-density lipoprotein (LDL) using the pathway described above (Anderson and Dietschy, 1978; Vahouny et al., 1985). In addition, the rat adrenal can use high-density lipoprotein (HDL) as a source of steroidogenic cholesterol by a pathway that involves a receptor—one that is not subjected to endocytosis (Gwynne and Hess, 1980; Gwynne and Strauss, 1982). The number of these HDL receptors is increased by ACTH (Gwynne and Hess, 1980; Gwynne and Strauss, 1982). To the extent that Y-1 mouse adrenal tumor cells reflect the behavior of normal mouse cells, this species uses the classical LDL pathway (Hall and Nakamura, 1979). Bovine adrenal cells do not store large amounts of cholesterol ester, but they employ the LDL pathway (Kovanen et al., 1979).

More than 40 years ago, C. N. H. Long pointed out that administration of ACTH to rats produces extensive depletion of adrenal cholesterol (Long, 1945). More than half of the total adrenal gland cholesterol is consumed within 12 hr of a single injection of ACTH. It was later shown that the disappearing cholesterol can be accounted for as corticosteroids (Péron and Koritz, 1960). It would appear that, *in vivo*, intense stimulation by ACTH causes depletion of cholesterol available for steroidogenesis, thereby challenging the mechanisms that supply such cholesterol to the cell. Since prolonged stress can cause high output of adrenal steroids for several days (Moore, 1957), we must conclude that the adrenal gland can obtain sufficient cholesterol to underwrite high levels of corticosteroids from two sources, namely synthesis from acetate and uptake of lipoprotein once the stores of cholesterol ester have been exhausted. It appears that under normal conditions *in vivo,* the rat and the human adrenal cortex contain sufficient stores of cholesterol ester to support acute responses to stress. These depots are derived from circulating lipoproteins from which cholesterol esters have been deesterified and reesterified as described above. This source accounts for at least 85% of the steroids secreted by the adrenal under basal or stimulated conditions (Vahouny *et al.*, 1985). A small amount of steroidogenic cholesterol is synthesized from acetate; this source of cholesterol may be important during prolonged stress or in those experimental conditions in which adrenal cells do not have access to lipoproteins. It should be pointed out that the fatty acid composition of adrenal cholesterol esters is most unusual in that it contains a high proportion of polyunsaturated fatty acids (Vahouny *et al.*, 1985). In this respect adrenal stores of cholesterol esters differ markedly in fatty acid content from the triglycerides and phospholipids of adrenal cells, and from circulating esterified cholesterol (Vahouny *et al.*, 1985). This last observation shows that when cholesterol ester in LDL is taken up by adrenal cells, it must be subjected to the sequence of deesterification and reesterification.

B. THE PATHWAY

The pathway to corticosteroids involves the conversion of cholesterol to cortisol:

4. ACTH AND CORTICOSTEROIDOGENESIS

The enzymatic reactions required for this conversion are as follows:

Activity	C atom
Hydroxylation	22, 20, 17α, 21, 11β
Dehydrogenation	3β
Isomerization	$\Delta^{4,5}$
C—C cleavage	20, 22

Hydroxylation and lyase activity require cytochromes P-450. Dehydrogenation at 3β requires a typical pyridine nucleotide dehydrogenase, and an isomerase moves the double bond to the more stable α,β-unsaturated ketone (Hall, 1984a, 1986). The pathway begins with conversion of cholesterol to pregnenolone (C_{27} side-chain cleavage):

CHOLESTEROL → (C_{27} SIDE-CHAIN CLEAVAGE) → PREGNENOLONE

It is generally agreed that 17α-hydroxylation precedes 21-hydroxylation in the synthesis of 17α-hydroxy-C_{21} steroids, and that the last step in the pathway is 11β-hydroxylation (Samuels, 1960; Hall, 1984a). These "rules" do not define the sequence of reactions between pregnenolone and 11-deoxycortisol. The most common sequence for these reactions is as follows:

Pregnenolone → [3β-OHSD, $\Delta^{4,5}$ KSI] → Progesterone → [17α-OHase] → 17α-OH Progesterone → [21-OHase] → 11-Deoxycortisol → [11β-OHase] → Cortisol

OHase: Hydroxylase
OHSD: Hydroxysteroid dehydrogenase
KSI: Ketosteroid isomerase

Alternative sequences are used in some species (Samuels, 1960; Hall *et al.*, 1964; Hall, 1986) but the basis for the relative use of different pathways is not clear. Before we consider the individual steps in the pathway, the subcellular distribution of the relevant enzymes must be considered. This can be summarized as follows:

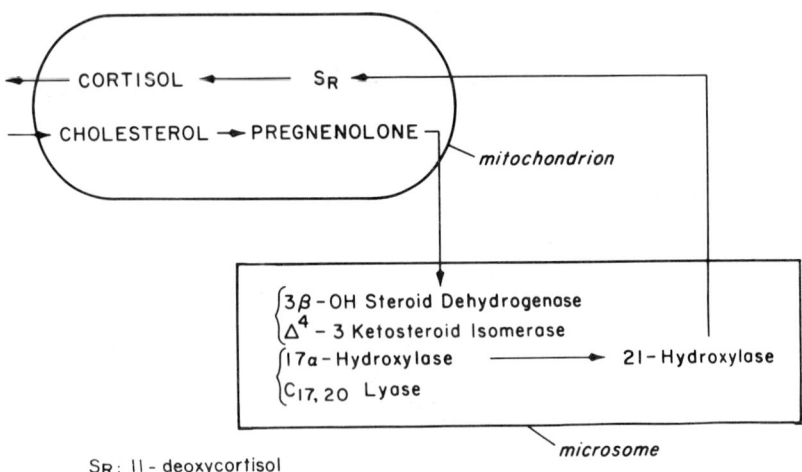

S_R: 11-deoxycortisol

The diagram reveals the movement of intermediates between mitochondria and microsomes which constitute important elements in the pathway.

1. C_{27} Side-Chain Cleavage

This enzyme is a cytochrome *P*-450 (Shikita and Hall, 1973a) that is situated on the inner aspect of the inner mitochondrial membrane (Yago *et al.*, 1970; Mitani *et al.*, 1982). It is responsible for catalyzing the conversion of cholesterol to pregnenolone in three steps:

The cleavage of the C—C bond is an unusual reaction for cytochrome *P*-450, but this step, like the other two, requires the heme moiety of the *P*-

450 (Hall et al., 1975); and like all cytochromes P-450, it shows the classical stoichiometry of a monooxygenase (Shikita and Hall, 1974):

20,22-Dihydroxycholesterol + NADPH + H$^+$ + O$_2$ → Pregnenolone + NADP$^+$ + H$_2$O
+ Isocapraldehyde

In aqueous media, the C_{27} side-chain cleavage enzyme associates with itself to give an active form consisting of 16 identical subunits (Shikita and Hall, 1973a). Like bacterial and other mitochondrial cytochromes P-450, the enzyme receives electrons from NADPH via two electron carriers, the iron–sulfur protein adrenodoxin and the flavoprotein adrenodoxin reductase.

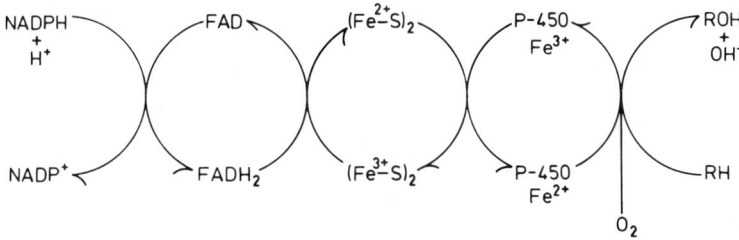

(Fe^{2+}—S)$_2$, reduced adrenodoxin; FADH$_2$, reduced adrenodoxin reductase

2. 3β-Hydroxysteroid Dehydrogenase–Δ^{4,5} Ketosteroid Isomerase

Pregnenolone formed from side-chain cleavage leaves the mitochondrion to be converted to progesterone by two microsomal enzymes which, although closely associated, appear to be distinct (Penning and Covey, 1982). An NAD$^+$-linked dehydrogenase leads to the formation of the Δ^5-3-ketone which spontaneously forms the Δ^4-3-ketone. However this conversion proceeds slowly at neutral pH and all steroid-forming organs possess an isomerase to catalyze the reaction (Samuels, 1960):

PREGNENOLONE → PROGESTERONE

3. 17α-Hydroxylase

Progesterone is converted to 17α-hydroxyprogesterone by an enzyme that closely resembles the $C_{17,20}$-lyase from testis that converts progesterone to androstenedione (Nakajin and Hall, 1981; Nakajin et al., 1981, 1983, 1984):

PROGESTERONE → [17α-OHASE] → 17α-OH PROGESTERONE → [$C_{17,20}$-Lyase] → ANDROSTENEDIONE

The adrenal enzyme catalyzes the same two reactions, although *in situ* the cleavage step occurs only to a limited degree. What regulates the degree to which lyase activity, and hence the synthesis of androgens, occurs *in vivo* is not clear since the adrenal and testicular enzymes are equally active in this respect *in vitro* (Nakajin and Hall, 1981; Nakajin *et al.*, 1984). The 17α-hydroxylase is a microsomal cytochrome *P*-450 that, like other such enzymes, receives electrons from a single electron carrier, the flavoprotein cytochrome *P*-450 reductase (Nakajin and Hall, 1981):

NADPH + H^+ → FMN-FAD → P-450 Fe^{2+} + O_2 + RH → Fe^{3+} P-450 → R-OH + OH^-
NADP$^+$ ← $FMNH_2$-$FADH_2$

4. 21-Hydroxylase

In the adrenal cortex most of the 17α-hydroxyprogesterone is not subjected to lyase activity but becomes substrate for 21-hydroxylase to produce 11-deoxycortisol:

17α-Hydroxyprogesterone → [21-OHASE] → 11-Deoxycortisol

It was pointed out above that it is generally believed that 17α-hydroxylation cannot occur after 21-hydroxylation (Eichhorn and Hechter, 1957). However, several species, including the rat, lack 17α-hydroxylase in the adrenal gland, so that 11-deoxycorticosterone (DOC) is the substrate for the final step of 11β-hydroxylation. In this case the secreted hormone is corticosterone rather than cortisol (Samuels, 1960). The 21-hydroxylase is a typical microsomal *P*-450 (Kaminami *et al.*, 1980; Yuan *et al.*, 1983).

5. *11β-Hydroxylase*

The pathway ends with the return of the steroid (11-deoxycortisol or DOC) to the inner mitochondrial membrane where a typical cytochrome *P*-450 inserts a hydroxyl group at the 11β-position (Watanuki *et al.*, 1977, 1978):

11-Deoxycortisol → [11β-OHase] → Cortisol

The cortisol (corticosterone in the case of the rat) is now ready to leave the cell as the secreted hormone.

These are the components of the steroidogenic pathway responsible for producing cortisol. This pathway is regulated so that in response to ACTH the output of cortisol can increase within a few minutes to levels more than 10 times higher than those seen in the resting, unstressed organism.

C. The Energy

Those reactions that require cytochrome *P*-450 use NADPH provided by mitochondria and microsomes while the 3β-dehydrogenase uses NAD^+. Other sources of energy required for the regulated movement of the intermediates through the cell have not so far been defined.

III. Stimulation of Steroid Synthesis

An advance in our understanding of metabolic activity has been the gradual acknowledgment that most cells possess only a limited repertoire of mechanisms by which they can respond to appropriate stimulation. What varies from cell to cell are the exact nature of the stimulus and the characteristic output of the cell. In spite of the best efforts by biologists who insist on isolating their investigations into arbitrary groupings called disciplines, cells use very similar regulatory mechanisms whether we consider the response of a nerve cell to a neurotransmitter, an endocrine gland to a hormone, or an exocrine cell to some local change. These responses involve proteins which may be synthesized *de novo* or modified posttranslationally. Two important messengers have been found to mediate these changes, namely cyclic AMP and Ca^{2+}. These are the tools available to all cells for the regulation of their functions in response to external stimuli. In what follows we will examine current evidence concerning the involvement of these changes in proteins produced by these messengers in the adrenal cell. Among the instruments within the cell that may participate in these responses, the following are important: (1) the plasma membrane in which an external change (increase in plasma ACTH) is expressed through the second messengers cyclic AMP and perhaps Ca^{2+}; (2) the nucleus which programs the nature and number of proteins being synthesized; (3) mitochondria which provide energy to support these responses; (4) microsomes where synthetic reactions occur; (5) Golgi apparatus in which refinements in the structures of proteins occur; (6) lysosomes in which various components of the cell are degraded; (7) the cytoplasm in which phosphorylation and innumerable other changes occur; and above all (8) the cytoskeleton which divides the cell into functional compartments and provides surfaces on which the enzymatic activities of cells take place in layers of bound water. The cytoskeleton deserves special mention because we understand it very incompletely and because it appears to pervade all cellular activities by organizing the internal structure of the cell (Hall, 1984b).

In addition to these mechanisms common to all cells, we must not forget the steroidogenic pathway itself which is specific to the cell in question. Here we can ask which step or steps in the synthesis of steroids is (are) altered by stimulation with ACTH. The answer to this question is obviously important when we consider the mechanism by which ACTH acts.

IV. Site of Action of ACTH in the Steroidogenic Pathway: Cholesterol Transport

The steroidogenic pathway is peculiar in that the substrate (cholesterol) is stored in the cytoplasm and yet the first enzyme in the pathway is situated in the inner mitochondrial membrane. When the mechanism of action of ACTH was first explored, biologists were not used to thinking about intracellular transport, so the problem of how cholesterol reaches mitochondria came to the attention of those working in the field as the result of experiments directed at other questions.

Early experiments showed that the side-chain cleavage of cholesterol is the slow step in the pathway (Stone and Hechter, 1954; Karaboyas and Koritz, 1965; Hall, 1967; Hall and Young, 1968). Three lines of evidence pointed to the supply of cholesterol to the mitochondrial P-450 as the principal factor regulating the rate of the side-chain cleavage reaction: (1) Garren and co-workers confirmed earlier studies (Ferguson, 1963) showing that the action of ACTH is inhibited by inhibitors of protein synthesis and, moreover, this inhibition affects steroidogenesis before the side-chain cleavage of cholesterol (Garren et al., 1965); it was proposed that protein synthesis is necessary for the transport of cholesterol to the mitochondrial enzyme and that this process is stimulated by ACTH. (2) Nakamura et al. (1980) showed that when the production of pregnenolone by isolated mitochondria from Y-1 cells was measured, mitochondria from cells treated with ACTH showed the same production of pregnenolone as those from untreated cells. However, if the cells were incubated with aminoglutethimide to inhibit the side-chain cleavage of cholesterol, with and without ACTH, mitochondria from cells treated with ACTH showed much greater production of pregnenolone than those from cells incubated without ACTH (Nakamura et al., 1980). Inhibition of side-chain cleavage allowed cholesterol to accumulate in the inner mitochondrial membrane. When mitochondria were prepared and washed to remove the inhibitor, the accumulated cholesterol was used for the production of pregnenolone (Nakamura et al., 1980). Without the inhibitor, removal of the mitochondria from the cytosol, with its stores of cholesterol, prevented the organelles from expressing the stimulating effect of ACTH (Hall, 1984b, 1985a). Moreover, the rate of side-chain cleavage must be fast relative to the rate of cholesterol transport. These studies also revealed the fact that entry of cholesterol into mitochondria requires some special mechanism because one could not increase the production of pregnenolone by isolated mitochondria by adding cholesterol to the isolated organelles (unpublished). (3) A number of studies have shown that the production of steroids by adrenal cells can be increased by providing an exogenous substrate for the C_{27} side-chain cleavage reaction in the form of hydroxycholesterol (22- or 25-hydroxy) (Mason et al., 1978; Bakker et al., 1979) or in the form of LDL (Faust et al., 1977; Carr and Simpson, 1981). It appears that the hydroxycholesterols can readily reach the mitochondrial enzyme, whereas cholesterol itself must enter a special, regulated pathway to the enzyme; this regulated pathway is slow. These studies plainly showed that the supply of cholesterol to the side-chain cleavage enzyme is important in the regulation of steroid production and in the mechanism of action of ACTH.

V. The Role of Cyclic AMP

The first major discovery concerning the mechanism of action of ACTH was the finding made by Haynes and co-workers that the hormone increases adrenal levels of cyclic AMP and that the cyclic nucleotide in turn stimulates the synthesis of adrenal steroids (Haynes, 1958; Haynes *et al.*, 1959). These observations established cyclic AMP as a second messenger for ACTH, and innumerable reports have confirmed this idea. The question of whether all the effects of ACTH can be accounted for by increased production of cyclic AMP remains unsettled. The binding of ACTH to its receptor is discussed in Section VII. Here we need only note that ACTH binds to one (Buckley and Ramachandran, 1981) or two (Yangibashi *et al.*, 1978) classes of receptor, and that this binding results in the production of cyclic AMP (Buckley and Ramachandran, 1981). A great excess of receptors for ACTH exist in adrenal cells because, when a fraction of the total population of receptors is occupied, maximal production of steroids is observed (Buckley and Ramachandran, 1981).

It is difficult to exclude the possibility that ACTH evokes responses in addition to increased production of cyclic AMP, responses which are necessary for increased production of steroids. For example, increased production of cyclic GMP has been proposed as an additional mechanism (Perchellet *et al.*, 1978). The best evidence for cyclic AMP as the only second messenger comes from the work of Schimmer *et al.* with mutant strains of Y-1 cells. Mutants with a normal receptor–cyclase system but with defective protein kinase show impairment of the steroidogenic response to ACTH (Schimmer, 1980). In the various mutants studied, the defects in protein kinase and in steroidogenesis were so related as to suggest that the kinase(s), and hence cyclic AMP, is (are) essential for the steroidogenic response to ACTH, and that the cyclic nucleotide is the only second messenger (Schimmer, 1980). This is the view that will be followed in this discussion, and the case for cyclic GMP will be considered provocative but unproven (Perchellet *et al.*, 1978). While cyclic AMP is essential for the action of ACTH, other agents may be involved.

Although cyclic AMP is the second messenger for ACTH, the role of the nucleotide may prove more complex than this statement indicates. It is, for example, not necessary to propose that the cell is flooded throughout with cyclic AMP in response to ACTH. It would be more likely, on the face of it, that the concentration of the nucleotide would rise in specific regions of the cell, although no information on this subject is yet available. It is known, however, that cyclic AMP leaves the adrenal cell (Schimmer and Zimmerman, 1976) and that the plasma membrane possesses binding proteins for the cyclic nucleotide (Wen et al., 1985). So far, studies of mitochondrial cyclic AMP have not been reported, although we have not found any response to this substance added to mitochondria or submitochondrial systems (unpublished). Several reports have suggested that mitochondrial cytochromes *P*-450 are phosphorylated under the influence of cyclic AMP (Caron et al., 1975; Vilgrain et al., 1984). If these findings are confirmed, the role of cyclic AMP as second messenger will take on new meaning. Finally, it is necessary to consider bound cyclic AMP as well as that released into the medium if we are to gain clearer insight into the role of this second messenger (Sala et al., 1979).

VI. The Role of Protein Synthesis

The steroidogenic actions of ACTH and cyclic AMP are inhibited by inhibitors of protein synthesis, e.g., puromycin (Ferguson, 1963) and cycloheximide (Garren et al., 1965). However, ACTH does not increase the incorporation of amino acids into total adrenal protein (unpublished). These findings suggest that cyclic AMP promotes the synthesis of a small number of proteins—too small to be detected by methods that measure total protein synthesis.

At the present time those investigators who work with adrenal cells do not appear to be impressed with evidence of nonspecific effects of cycloheximide, i.e., effects that cannot be directly attributed to inhibition of protein synthesis. Such effects include inhibition of the side-chain cleavage of cholesterol by interference with electron transport (Hsu and Kimura, 1982). In fact, inhibition of any response to ACTH by cycloheximide is currently regarded as a *sine qua non* for involvement of the response in the mechanism of action of ACTH. It is undoubtedly prudent to use puromycin as well as cycloheximide and to determine ED_{50} for inhibition of protein synthesis and that of the response in question. Such simple procedures would make the use of cycloheximide more convincing. Moreover, ACTH stimulates some changes in adrenal cells that do not require protein synthesis (see below).

Nevertheless, new protein(s) is (are) required for the response to ACTH. The production of these proteins is rapid—one or more of the new proteins has (or have) short half-lives of the order of 2–6 min (Lowry and McMartin, 1974; Schulster et al., 1974). These proteins are evidently essential for increase in the production of steroids in response to ACTH although, as we will see, part of this response does not require new protein(s).

The difficulty associated with the synthesis of specific proteins is that such findings must remain purely descriptive in the absence of any knowledge of the functions of the proteins in question. Each of these reports implicating specific proteins reaches a dead end because, even if such a protein could be purified, the examination of the effects of antibodies to the protein on the response to ACTH would not only be technically difficult but something of a shot in the dark—in most cases, it does not prove possible to learn about the function of a specific protein by indirect associations. We will consider below possible ways of overcoming such an impasse by efforts to relate the protein to functions used in the response to ACTH. The properties of individual proteins synthesized under the influence of ACTH will also be considered.

VII. The Role of Phosphorylation

It is generally believed that the only biological action of cyclic AMP is exerted by way of protein kinase enzymes and is hence entirely the result of phosphorylation of proteins (Kuo and Greengard, 1969). The phosphorylated proteins are capable of altered function. In view of the role of cyclic AMP as the second messenger for ACTH, it is obvious that considerable importance is attached to phosphorylation of adrenal cell proteins in response to ACTH. Unfortunately, progress in this aspect of the mechanism of action of ACTH is severely limited by our present lack of insight into the nature of the altered function(s) of specific proteins following phosphorylation. Nevertheless, interest in this problem is stimulated by the advances in our understanding of the cascade of reactions involved in the activation of glycogen phosphorylase in which the involvement of phosphorylation of specific proteins and the concomitant changes in their activities are clearer.

Studies by Koroscil and Gallant (1980) revealed that a large number of proteins in all the major subcellular fractions of adrenal cells are phosphorylated under the influence of ACTH. Unfortunately, neither the proteins nor their functions can be identified by this approach. Podesta et al. (1979) have studies the phosphorylation of a specific protein in response to ACTH. This response was studied in some detail, and this phosphoprotein seems likely to be important in the response to ACTH. Again, however, these findings must wait on new approaches before the significance of this phosphorylation can be clarified. Another approach to this problem has been the examination of phosphorylation of proteins in specific cellular compartments in which the possibilities of functional roles of the proteins concerned might be examined. For example, studies with insulin revealed phosphorylation of proteins in the plasma membrane which could be related to glucose transport (Chang et al., 1974). This approach was used with ACTH (Widmaier et al., 1985). Unfortunately the problems involved in preparing highly purified plasma membranes have simply been ignored in many studies in spite of excellent reviews in which these problems are discussed (for example, Warren et al., 1966; Warren and Glick, 1969). Journals that are properly insistent on good evidence for the purity of organelles and molecules freely publish studies that purport to deal with plasma membrane when it is clear to an informed reader that these preparations are heavily contaminated by internal membranes. Studies of the phosphorylation of proteins in such preparations can only add to the present confusion. Widmaier et al. (1985), using highly purified plasma membranes (Osawa and Hall, 1985) from Y-1 cells, showed that three proteins are phosphorylated by cyclic AMP (270K, 35K, and 17K). The phosphorylation of these proteins is likely to be important in the response to ACTH, but so far no function can be associated with these proteins (Widmaier et al., 1985).

The cytoskeleton of Y-1 adrenal tumor cells has also been examined for phosphorylation by cyclic AMP. The cytoskeleton contains tightly bound protein kinase enzyme(s), of which one or more promote phosphorylation of two proteins (Osawa and Hall, 1985), increased phosphorylation of five proteins, and dephosphorylation of one cytoskeletal protein (Osawa and Hall, 1985). Again the functional significance of these changes is not clear but must presumably be related to alterations in the functions of the cytoskeleton that result from the action of ACTH.

Finally, at least one mitochondrial protein shows increased phosphorylation under the influence of ACTH (Bhargava et al., 1978).

VIII. The Role of Ca^{2+}

Investigation of the role of Ca^{2+} in the response to ACTH began when Birmingham et al. (1953) showed that Ca^{2+} is necessary for the steroidogenic response to ACTH *in vitro,* which is hardly surprising because Ca^{2+} is probably necessary for many or most biological activities. This line of investigation has proceeded by fits and starts according to advances in methodology. Little can be learned from the use of powerful chelating agents or ionophores, which proved confusing because drastic alteration of the levels of Ca^{2+} produced by these agents must cause innumerable nonspecific effects (Nakamura and Hall, 1978) that result from changes in which Ca^{2+} is necessary or permissive—changes which cannot be distinguished from involvement of this ubiquitous cation as a specific regulatory mechanism. More than thirty years later it has not been decided to everyone's satisfaction whether or not Ca^{2+} is required for the effects of the second messenger, cyclic AMP. Indeed a recent report proposed that the role of Ca^{2+} can be largely accounted for by its involvement in the interaction between ACTH and its receptor (Cheitlin et al., 1985), (Chapter 7). However, other studies show that the effect of Ca^{2+} not accounted for by this interaction may be extremely important in the regulation of steroid synthesis. The problem is at an intriguing stage, but the reader must be warned that no single account of the available data can be formulated to give a unifying hypothesis.

In the first place, Iida et al. (1985) using quin 2 and fura 2 saw no change in intracellular Ca^{2+} in Y-1 or bovine adrenocortical cells following addition of ACTH. While these observations make it unlikely that rapid influx of Ca^{2+} on the scale seen with responses to other hormones (e.g., epinephrine and vasopressin) (Michell, 1983; Williamson et al., 1985) occurs in the adrenal cortex with ACTH, they do not exclude other responses involving Ca^{2+}, e.g., faster, small responses and, above all, internal redistribution of Ca^{2+} in response to ACTH.

An elegant study by Yanagibashi has raised important possibilities for the role of Ca^{2+} in the control of steroid synthesis (Yanagibashi, 1979). Rat adrenal cells do not show uptake of Ca^{2+} until they are treated with ACTH which promotes influx of Ca^{2+} (Yanagibashi, 1979); this influx does not result from the action of cyclic AMP. Bovine cells, on the other hand, show uptake of Ca^{2+} without ACTH. This influx is accompanied by increased production of steroids; this response is inhibited by cycloheximide (Yanagibashi, 1979). Coming upon these findings from outside the problem of ACTH, one would say that the two types of cells, being prepared by different methods, are not strictly comparable. There are, however, many differences between rat and bovine adrenal cells, so that a true species difference in the role of Ca^{2+} in response to ACTH is entirely possible. Moreover, variations in external Ca^{2+} between 0.3 and 2.5 mM are not normally encountered by the cell, so that these experiments may cause an unphysiological uptake of Ca^{2+} that reveals the effect of an internal redistribution of the cation. The findings of Yanagibashi force us to consider that Ca^{2+} may serve as a second messenger for ACTH (Yanagibashi, 1979). It should be added that electrophysiological studies of the adrenal cell have not proved rewarding (Mathews and Saffran, 1973). There is nothing to suggest that hypopolarization of the adrenal cell is a feature of the stimulation of steroid synthesis by ACTH. A recent study has reported K^+-dependent Ca^{2+} channels in plasma membrane of rat adrenal that increase in number on treatment with ACTH (Kenyon et al., 1985). This interesting observation deserves further study and may lead to a better understanding of the role of Ca^{2+} in the response to ACTH.

Closely related to the biology of Ca^{2+} is the Ca^{2+}-binding protein calmodulin (Means and Dedman, 1980; Cheung, 1980). The steroidogenic responses to ACTH and cyclic AMP are inhibited by trifluoperazine (Hall et al., 1981b). Moreover, injection of calmodulin into Y-1 cells by fusion with liposomes stimulates steroid synthesis and the transport of cholesterol to mitochondria (Hall et al., 1981a). Since it is known that most effects of calmodulin are explained without new synthesis of this protein (Means and Dedman, 1980), these findings suggest that ACTH may cause redistribution of calmodulin within the cell. The injected calmodulin presumably enters compartments in the cell from which it is partly excluded in the resting condition. Calmodulin injected into Y-1 cells without bound Ca^{2+} was much less effective in stimulating steroid synthesis (Hall et al., 1981b). Presumably, the injected calmodulin–Ca^{2+} reaches parts of the cell in concentrations that evoke a response. Similar changes could occur in the intact cell *in vivo* in response to cyclic AMP which could, in turn, regulate the distribution of the protein within the cell.

IX. The Possible Role of Protein Kinase C

The discovery of a protein kinase that is dependent upon Ca^{2+} and phospholipid for activity (so-called protein kinase C) (Nishizuka, 1984; Nishizuka et al., 1984) has made this kinase a possible candidate for a role in many cellular activities. Already this kinase has come to occupy a central role in numerous functions in a wide variety of cells (Nishizuka, 1984; Nishizuka et al., 1984). Protein kinase C may be a receptor for certain tumor-promoting agents called phorbol esters (Kikkawa et al., 1983; Niedel et al., 1983), and it may be a receptor for diacylglycerol (Kikkawa et al., 1983; Niedel et al., 1983). Since phorbol esters have something in common with diacylglycerol in structure, the idea has developed that the physiological regulation of the kinase is by way of diacylglycerol released from cellular phospholipids by phospholipase C. If an external stimulus activates the lipase, diacylglycerol would stimulate protein kinase C which would, in turn, promote phosphorylation of specific substrates with the changes in function of these proteins that result from phosphorylation. Whereas the half-life of diacylglycerol is limited by cellular metabolism, the action of the tumor promoters would not be subject to such regulation, thereby producing excessive and unregulated activity of protein kinase C which could lead to the production of tumors.

Since protein kinase C is Ca^{2+}-dependent, an increase in the intracellular concentration of this cation could activate the kinase which would phosphorylate its substrates. In this hypothesis, protein kinase C would be to Ca^{2+} what cyclic AMP-dependent protein kinase (PKA) is to cyclic AMP:

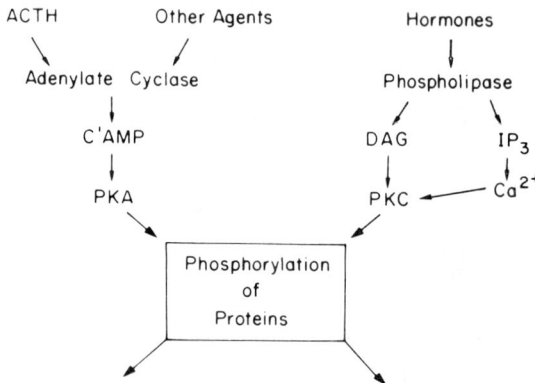

The diagram is intended to suggest that several stimuli beside ACTH on the one hand (e.g., cholera toxin) and diacylglycerol on the other hand are capable of activating the two phosphorylating systems. Arrows diverging from this common mechanism imply that different substrates may be phosphorylated or that phosphorylation of a common substrate may occur at different amino acid residues (Vulliet et al., 1985). Such an hypothesis would provide an explanation for different responses of one cell to different blood-borne stimuli. The theoretical difficulty with these ideas, in the case of the adrenal, lies in the fact that there is no known physiological stimulus for the regulation of steroid synthesis apart from ACTH. Moreover, the cyclic AMP pathway is well established, whereas the Ca^{2+} pathway remains uncertain.

It is clear that bovine adrenal cells, Y-1 cells, and rat fasciculata cells all possess a kinase activity that meets the criteria for protein kinase C (Widmaier and Hall, 1985b; Vilgrain et al., 1984). Phorbol ester stimulates the activity of the enzyme and also stimulates production of steroids by these adrenal cells (Widmaier and Hall, 1985b). However, in Y-1 cells the steroidogenic response to phorbol ester is much smaller than that to ACTH. Moreover, cycloheximide and puromycin increase the activity of protein kinase C (PKC) in adrenal cells (Widmaier and Hall, 1985b). These considerations made it unlikely that PKC is involved in this response. In bovine adrenal it was reported that ACTH causes movement of protein kinase C from membranes to cytosol (Vilgrain et al., 1984), whereas movement in the opposite direction is seen with other stimulating

4. ACTH AND CORTICOSTEROIDOGENESIS

agents in other cells (Hirata *et al.*, 1985). Finally, it has been reported that protein kinase C is responsible for phosphorylation of the mitochondrial *P*-450 that catalyzes C_{27} side-chain cleavage (Vilgrain *et al.*, 1984). These observations suggest that protein kinase C cannot be a component of the response to ACTH in the universal way that cyclic AMP serves in that capacity. In that case adrenal protein kinase C remains an enzyme in search of a function so far as the adrenal cortex is concerned. It is too early to assess such possibilities.

X. The Possible Role of Phospholipids

During the past 8 years, a new theme in the regulation of numerous cell functions has emerged in which phosphoinositide metabolism occupies the central role. It has become clear that phosphatidylinositides engage in a cycle of synthesis and breakdown (Michell, 1983; Williamson *et al.*, 1985). From phosphatidylinositol (PI), a branch cycle occurs in which the inositol moiety is phosphorylated successively at positions 4 and 5 of the ring to give rise to the so-called polyphosphoinositides abbreviated PIP and PIP_2:

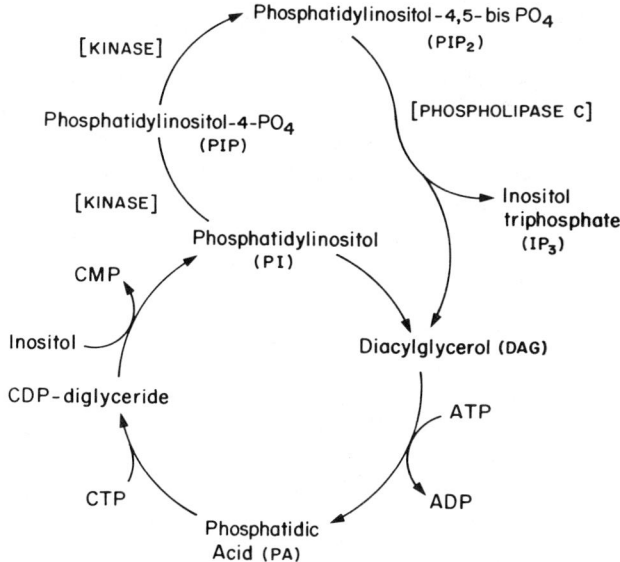

The breakdown of PIP_2, catalyzed by phospholipase C, gives rise to two important products, namely diacylglycerol and inositol triphosphate:

It is proposed that a variety of hormones and neurotransmitters bind to specific receptors in the plasma membranes of target cells and one consequence of this binding is activation of phospholipase C. It was pointed out above that diacylglycerol activates protein kinase C so that the proposed mechanism leads to phosphorylation of substrates for this enzyme. Meanwhile, inositol triphosphate has been shown to cause redistribution of intracellular Ca^{2+}. The hexose triphosphate can serve as a Ca^{2+} ionophore (Downes and Michell, 1981; Berridge and Irvine, 1984) which could promote either entry of Ca^{2+} into the cell or redistribution of internal Ca^{2+} by mobilizing Ca^{2+} from mitochondria or endoplasmic reticulum. Although it is too early to generalize, it appears that inositol triphosphate stimulates release of Ca^{2+} from endoplasmic reticulum and that entry of Ca^{2+} via the plasma membrane or from mitochondria are less important. It also seems clear that phospholipase C can be activated without mobilization of Ca^{2+}; that is, the enzyme requires Ca^{2+} for activation but resting levels of Ca^{2+} are sufficient for such activation so that mobilization of Ca^{2+} is a consequence of activation of phospholipase C and not a requirement (Berridge et al., 1983; Thomas et al., 1984; Charest et al., 1985).

These consequences of the activation of phospholipase C could explain a vast range of cellular responses since they include increase in intracellular Ca^{2+} and activation of protein kinase C. Since cytoplasmic Ca^{2+} in most cells is normally maintained at submicromolar concentrations, and since plasma Ca^{2+} and internal depots of Ca^{2+} are in the micromolar range, inositol triphosphate could disturb this balance and lead to activation of any Ca^{2+}-dependent mechanism. Such a system could be seen as providing the cell with a second messenger pathway in addition to, or instead of, the much better understood cyclic AMP pathway. It was pointed out above that we are inhibited about the interpretation of such a pathway by the deeply entrenched notions that ACTH provides the only physiological regulation of steroid production and that cyclic AMP accounts for all of the effects of ACTH on adrenal cells.

With this background there appeared a series of papers from one laboratory describing the phosphoinositide pathway in rat adrenal cells and activation of the pathway by ACTH (Farese, 1983, 1984; Farese et al., 1979, 1980). The situation is complicated by the fact that these workers reported changes in the amount or mass of the phosphatidylinositols (Farese, 1983; Farese et al., 1979, 1980). Most other workers have relied upon ^{32}P-labeled or [^{3}H]*myo*-inositol to label the relevant pathways before stimulation. It is true that, working *in vivo* with the rat, it is difficult to be sure that ^{32}P-labeled inorganic phosphate has labeled the ATP and phospholipid pools to steady state before administration of ACTH. These workers have not reported data with [^{3}H]*myo*-inositol. However, the sensitivity of the method for measuring the mass of the various components of the pathway greatly limits confidence in the findings. Without doubt, cultured cells provide a more flexible system. One difficulty with cell culture lies in the loss of some features of the fully differentiated cell as it occurs in the intact organism. However, bovine adrenal cells and mouse adrenal tumor cells show excellent steroidogenic responses to ACTH and cyclic AMP. If the phosphatidylinositol mechanism is an essential component of the steroidogenic response to ACTH, it should be possible to see such changes in these cultured cells. Moreover, cell culture makes it easy to use isotopic precursors which increase the sensitivity of the method; this would enable the investigator to compare his findings with those of workers using other systems. It was pointed out above that Iida et al. (1985) observed no change in intracellular Ca^{2+} in response to ACTH with either quin 2 or fura 2. Moreover, these workers saw no change in incorporation of either ^{32}P-labeled or [^{3}H]*myo*-inositol into the relevant intermediates (Iida et al., 1985) in either bovine fasciculata or Y-1 cells. By contrast, Iida et al. (1985) clearly showed that angiotensin II produced a rapid decrease in the level of ^{32}P in polyphosphoinositides and an increase in intracellular Ca^{2+} in bovine glomerulosa cells (Iida et al., 1985). These results confirmed findings by other workers in glomerulosa cells (Farese et al., 1984) and demonstrated that the negative findings of Iida et al. (1985) with ACTH could not be attributed to failure of the methods used. It must therefore be concluded that either the phosphatidylinositol–Ca^{2+} mechanism is not a universal requirement for the steroidogenic action of ACTH or that the change in phosphoinositides exerts different effects from those seen in response to other stimulating agents that use the phosphoinositol mechanism with the exception of LH which the same group has studied (Farese et al., 1984). Their findings with LH are essentially the same as those discussed above for ACTH (Farese, 1984). It should also be pointed out that the responses to ACTH and LH reported by Farese *et al.* are qualitatively different from those seen with other agents employing the phosphoinositide pathway. The usual response is a very rapid (<1 min) fall in the level of polyphosphoinositides. The responses to ACTH and LH reported by Farese *et al.* involve increases in the synthesis of phosphatidylinositol itself.

These considerations place a high premium on the ability of an independent group to confirm these findings in the rat. The point is crucial to everyone working on the steroidogenic actions of ACTH and LH because, if the idea is correct, all our views about the role of cyclic AMP in the response to ACTH must change to accommodate this separate mechanism. If it is not correct, much valuable time will be lost in the pursuit of a false lead.

Less spectacular changes in the phospholipids of mitochondria in response to ACTH have been reported (Igarashi and Kimura, 1984). Such changes may influence the activities of the steroidogenic enzymes in the inner mitochondrial membrane.

XI. The Roles of Subcellular Components of the Adrenal Cell

Reference has already been made to the involvement of different parts of the adrenal cell in the response to ACTH, e.g., plasma membrane, mitochondria, etc. However, the emphasis of the preceding discussion has been on molecular mechanisms with less concern for the site or sites within the cell at which these mechanisms occur. We must now consider the response of structures within the cell to ACTH.

A. THE CYTOSKELTON

It was common during the 1960s to hear biologists say that the cell is more than a bag of enzymes. This expression was intended to convey the idea that we do not find a useful model for a cell in the concept of cytoplasm as a uniform, well-stirred aqueous solution. The idea is too fundamental to require documentation at the present time. What was not clear during the 1960s was the basis for the nonhomogeneity of the cytoplasm. We now know that the cytoskeleton is the major factor in producing compartments within the cytoplasm. Microtubules and microfilaments break the cytoplasm into gross compartments and, since these structures can be constructed and disassembled under controlled conditions (Korn, 1978; Olmsted and Borisy, 1973), compartments within the cytoplasm can be made to order as the needs of the cell change. Moreover, the state of cytoplasm—the degree to which it can be described as sol or gel—is determined by the degree of cross-linking of microfilaments (Taylor and Condeelis, 1979). The general shape of the cell is maintained by intermediate filaments (Lazarides, 1980). In addition to these elements of the cytoplasm, there appears to be a reticular system that subdivides the cytoplasm into still smaller compartments (Wolosewick and Porter, 1979). Unfortunately, our understanding of these finer subdivisions of the cytosol is limited by technical problems associated with interpretation of electron micrographs (Ris, 1985). The important contribution of Ris casts some doubt on the existence of microtrabeculae as originally described (Ris, 1985). In any event it is clear that the properties of the cytoplasm are greatly influenced by compartmentation produced by the cytoskeleton. Furthermore, the most important influence of the cytoskeleton lies in the surfaces which it provides because these surfaces bind cell water and alter its properties (Clegg, 1984). By limiting the freedom of bound water, the cytoskeleton appears to organize cellular proteins in relation to specific surfaces. The fact that the enzymes of glycolysis are bound to microtubules (Ottaway and Mowbray, 1977) encourages the idea that biosynthetic pathways may be organized in the bound water associated with the surfaces of microtubles. This would explain, at least in part, how the cell manages to accomplish at 37°C and atmospheric pressure what takes the chemist extreme conditions to achieve. Biosynthetic pathways must depend on something more efficient than random encounters between reacting species. We are only beginning to find ways of exploring these important aspects of cell biology.

The first evidence that the cytoskeleton may be involved in the response to ACTH came from the observation that ACTH and cyclic AMP cause Y-1 cells to round up (Yasamura, 1968). This drastic change in shape must involve a radical reorganization of the cytoskeleton. It is true that one cannot readily relate rounding of a cultured cell to what may happen in an organ *in vivo*. Nevertheless, ACTH and cyclic AMP are capable of altering the cytoskeleton of adrenal cells.

It was subsequently shown that cytochalasin B inhibits the steroidogenic responses to ACTH and cyclic AMP (Mrotek and Hall, 1975, 1977). These studies were accompanied by important controls which made a nonspecific effect of cytochalasin unlikely (Mrotek and Hall, 1975, 1977). Moreover, the inhibitory effect is localized to the transport of cholesterol to mitochondria (Mrotek and Hall, 1975, 1977). The action of cytochalasin is rapid in onset and freely reversible (Mrotek and Hall, 1977). Doubts about the specificity of the action of cytochalasin were diminished by showing that the inhibitory effects of cytochalasins B, D, E, and reduced B could be related to the relative binding of each cytochalasin to adrenal actin (Hall *et al.*, 1981a), strongly suggesting that inhibition of the steroidogenic response results from binding to actin rather than the result of a nonspecific effect (Hall *et al.*, 1981a). The involvement of actin in the response to ACTH was further demonstrated by injection of antiactin into Y-1 cells following fusion with liposomes containing the antibody (Hall *et al.*, 1979). It was observed that antiactin inhibits the responses of Y-1 cells to ACTH and that inhibition can be prevented by treating the antibody with excess actin (Hall *et al.*, 1979). These findings set to rest problems regarding the specificity of the action of cytochalasins. However, they left three important questions unanswered, namely, where in the cell does actin act, how much antibody relative to cell actin is required for inhibition, and how is the antibody acting.

These questions were approached by means of a new inhibitor and a new mode of injection. DNase I can be obtained in highly purified form, and its mechanism of action is well understood—it binds G-actin and prevents it from polymerizing (Blickstad *et al.*, 1978). Red cell ghosts provide an efficient and reproducible system for injecting macromolecules into cells (Schlegel and Reichsteiner, 1975). Moreover, the red cell ghosts can be subjected to lysis following the injection process so that any surface material that has not been injected can be removed. This means that the amount of a labeled macromolecule injected into the cell can be determined, e.g., by counting radioactivity in the treated cells.

With this new system (using fluorescent DNase I), it was found that DNase I appears in the cytoplasm but is excluded from mitochondria and nuclei (Osawa *et al.*, 1984). Moreover, DNase I inhibits the steroidogenic responses to ACTH and cyclic AMP at the step of the transport of cholesterol to mitochondria (Osawa *et al.*, 1984). Since DNase I is excluded from mitochondria, and since mitochondria from Y-1 cells do not contain actin (unpublished), we can conclude that DNase I inhibits transport of cholesterol to mitochondria as opposed to within mitochondria. However, this does not exclude an additional effect of ACTH on the intramitochondrial processing of steroidogenic cholesterol.

The studies with DNase I also showed that 50% inhibition of the response to ACTH is produced by 3×10^7 molecules of DNase per cell. The content of G-actin in the cell is 1.5×10^7 molecules per cell (Osawa *et al.*, 1984). Since DNase I acts by binding to monomeric or G-actin, it is clear that the action of ACTH on Y-1 cells requires a pool of G-actin which can undergo polymerization under the influence of ACTH.

The cytoskeleton is likely to differ from cell to cell so that generalizations concerning the involvement of microtubules and microfilaments must be made with caution. The response to ACTH in Y-1 cells is not inhibited by colchicine at concentrations that cause gross destruction of microtubules (Mrotek and Hall, 1978). On the other hand, the response of rat adrenals to ACTH is inhibited by cytochalasin B and colchicine (Crivello and Jefcoate, 1978), and that of bovine fasciculata cells is inhibited by colchicine but not by cytochalasin (Rainey et al., 1984). Cells vary considerably in shape and in cytoskeletal organization; these differences pose a variety of problems for the regulation of intracellular transport. It is therefore not surprising that all cells do not employ the same components of the cytoskeleton to direct intracellular traffic. These differences will only be explained when we know more about exactly what changes occur in the cytoskeleton in response to ACTH in the various cells.

B. The Mitochondrion

Since the mitochondria provide much of the energy used by a cell, the knowledge that side-chain cleavage takes place in mitochondria stimulated a search for an effect of ACTH on energy production by mitochondria. The results were interesting, but the case for such an effect of ACTH or cyclic AMP was not sustained. It turned out that C_{27} side-chain cleavage can be supported by reversed electron transport, i.e., reduction of NAD^+ by succinate (Hall, 1967). The NADH is converted to NADPH by transhydrogenation (Hall, 1972). However, there was no evidence that ACTH or cyclic AMP alters the production of NADPH.

A new look at the role of mitochondria in the response to ACTH was provoked by the observation that ACTH increases the transport of cholesterol to the inner mitochondrial membrane (Nakamura et al., 1980) and that this cholesterol leaks out of the inner membrane when mitochondria are kept at 4°C. It was observed that ACTH increases the high spin signal from adrenal mitochondria (Jefcoate and Orme-Johnson, 1975). When any cytochrome P-450 binds substrate, the Soret peak characteristic of the heme moiety of the enzyme shifts from 420 to 390 nm. This shift results from movement of the heme iron out of the plane of the ring system. This, in turn, is associated with a reorganization of the d orbital electrons to the so-called high spin form (Jefcoate and Orme-Johnson, 1975). This change provides a means of measuring the proportion of the enzyme that is in the enzyme–substrate complex (ES). It was found that, following treatment with ACTH, the proportion of the enzyme in the form of ES increased (Jefcoate and Orme-Johnson, 1975). It was subsequently shown that this change results from anaerobiosis of mitochondria which occurs, during preparation of the organelles, unless great care is taken. This high spin signal is not seen in aerobic mitochondria (Bell and Harding, 1974). It should be pointed out that the blood supply to the adrenal gland is the highest per unit weight of any organ in the body except the carotid body.

It is, therefore, very easy for mitochondria to become anaerobic when this vigorous blood flow is stopped. These studies did show, however, that ACTH increases the amount of steroidogenic cholesterol in adrenal mitochondria which was in agreement with studies using aminoglutethimide as discussed earlier (Nakamura et al., 1980). Anaerobiosis also inhibits side-chain cleavage, and when ACTH is administered the concentration of cholesterol in the inner membrane increases and the side-chain cleavage enzyme becomes loaded with substrate. These changes can be measured chemically (Nakamura et al., 1980) or by the high spin signal (Jefcoate and Orme-Johnson, 1975), whether inhibition of side-chain cleavage results from aminoglutethimide or anaerobiosis, and whether anaerobiosis is inadvertent (Jefcoate and Orme-Johnson, 1975) or deliberate (Nakamura et al., 1980). Moreover, the accumulated cholesterol can be converted to pregnenolone by incubating mitochondria after removing the inhibitor (aminoglutethimide) or introducing oxygen (Nakamura et al., 1980; Hall, 1985).

The inner and outer mitochondrial membranes were then studied in greater detail. Ohno et al. (1983) showed that the cholesterol content in the outer membrane increases on administration of ACTH and cycloheximide (Ohno et al., 1983). This finding was confirmed by Privale et al. (1983). It is clear that the step that is sensitive to cycloheximide is the transport of cholesterol from the outer to the inner mitochondrial membrane. Evidently ACTH stimulates the transport of cholesterol at two points in the pathway—to the outer membrane (Mrotek and Hall, 1977; Hall et al., 1979; Osawa et al., 1984) and from the outer to the inner membrane (Ohno et al., 1983; Privale et al., 1983). It has also been reported that a cholesterol carrier protein appears in greater amounts in the inner mitochondrial membrane after administration of ACTH (Connely et al., 1984). The evidence for the identity of this protein depends on one dimensional SDS gels, and the detailed characterization of the antibody used has not been given (Connely et al., 1984).

It is not clear how a carrier protein could accelerate the passage of cholesterol from outer to inner membrane, although it is known that more polar derivatives of cholesterol, e.g., 25-hydroxycholesterol, are metabolized by adrenal mitochondria much more rapidly than cholesterol itself (Mason et al., 1978; Bakker et al., 1979). A novel approach to this question was used by Lambeth et al. who measured the volume of intermembrane space in control and stressed adrenal mitochondria using a method based upon the distribution of [^{14}C]sucrose (Palmieri and Klingenberg, 1979). The volume of the space was decreased by stress (ACTH), and this change was inhibited by cycloheximide (Lambeth and Stevens, 1984). The authors propose that the decrease in intermembrane space results from enhanced contact between the two surrounding membranes. It is difficult to determine the validity of the method used. Presumably such changes should be visible with electron microscopy.

The events in the inner membrane associated with the side-chain cleavage of cholesterol have been studied in detail by Lambeth et al. (1982). Spectroscopic studies revealed that the two electron carriers (adrenodoxin and adrenodoxin reductase) and P-450 do not form a ternary complex. On the contrary, oxidized adrenodoxin forms a binary complex with reduced adrenodoxin reductase. When adrenodoxin is reduced it dissociates from this complex and binds to P-450. The formation of this second binary complex is promoted by binding of cholesterol to P-450 (Lambeth et al., 1982). Moustafa and Koritz (1977) showed that Ca^{2+} promotes dissociation of adrenodoxin from the first binary complex. These studies appear to have identified a Ca^{2+}-sensitive event in the inner mitochondrial membrane which could prove important in understanding the role of this ion in the regulation of steroid synthesis.

C. The Cytoplasm

A number of important events essential to the response to ACTH takes place in the cytoplasm.

1. *Cholesterol Ester Hydrolase*

ACTH stimulates the activity of a cholesterol esterase that is important in mobilizing cholesterol for steroid synthesis (Trzeciak and Boyd, 1974). Presumably, this step is essential to permit cholesterol to enter the steroidogenic pathway. The responding enzyme is a neutral cholesterol ester hydrolase that is phosphorylated by a cyclic AMP-dependent protein kinase. Phosphorylation increases the activity of the enzyme (Beckett and Boyd, 1977; Naghshinek et al., 1978).

2. *Sterol Carrier Protein*

The cytosol of adrenal cells that have been stimulated by ACTH stimulates the production of pregnenolone by isolated adrenal mitochondria (Vahouny et al., 1985). This stimulation is abolished by treatment with an antibody to sterol carrier protein 2 (SCP_2) isolated by Scallen and co-workers (Chanderbhan et al., 1982). It appears that SCP_2 facilitates transport of free cholesterol from lipid droplets to mitochondria so that, in concert with cholesterol ester hydrolase, it is responsible for mobilizing steroidogenic cholesterol from the lipid droplet stores of cholesterol ester (Chanderbhan et al., 1982). In addition, SCP_2 appears to facilitate transport of cholesterol through the mitochondrion to the side-chain cleavage enzyme (Vahouny et al., 1983).

3. New Proteins

More than five groups have reported the synthesis of specific proteins under the influence of ACTH. Unfortunately, each of these proteins has its own experimental setting so that it is not possible to relate one to the other. We must begin by considering the timing of these events. A steroidogenic response to ACTH has been reported within 1 min of addition of the hormone (Widmaier et al., 1985). We are justified in taking the earliest reported response since slower onset of a response is likely to be attributed to experimental conditions. If the average time required to translate eukaryotic mRNA is taken as 2 min (Vaughan et al., 1971), we may reasonably ask whether the synthesis of new protein can be necessary for this response. To answer this question we must take sides in the issue of whether or not the adrenal cell stores formed secretory products (cortisol and corticosterone). It has generally been assumed that no such stored material is present in the cell (Cam and Bassett, 1983; Mathew et al., 1985), and in any case there is no obvious break in the early time course (<2 min) of the response to ACTH (Widmaier et al., 1985). That is to say, the increased production of steroids on addition of ACTH results from increased synthesis of steroid from cholesterol rather than release of preformed hormone followed by increased synthesis; such a double response would be expected to show a complex relationship with time. If we accept this view, new protein cannot reasonably be invoked in the early response because we must include not only the 2 min required to synthesize the protein but the time required for it to reach the mitochondria and the time required for the synthesis of steroids. It is clear that no response would be possible for more than 2 min. Unfortunately, such arguments by exclusion are not persuasive. In the first place, we cannot exclude all possibility of internal stores of steroid hormone—some authors have proposed that the adrenal cell does, in fact, store glucocorticoids ready for secretion (Cam and Bassett, 1983; Mathew et al., 1985). In the second place, a break in the time curve could be difficult to detect. In fact, we cannot assume that all cells in a culture dish or an organ respond uniformly—it seems more probable that we are looking at the net response of a heterogeneous conglomeration of cells. It is here that work with single cells at temperatures that slow the various steps in the response will be important. We must reluctantly admit that the first stirrings of the response to ACTH cannot be described in detail and that the synthesis of new protein may or may not be necessary. Faced with these doubts we can do no more than consider each response on its own merits.

a. Y-1 Cells. Following treatment with ACTH, Nakamura and Hall (1978) observed two proteins (26K and 13K) that were synthesized on cytoplasmic ribosomes and transported to mitochondria (Nakamura and Hall, 1978). The time course of these events greatly exceeded the 2–6 min in which steroidogenic protein is believed to be synthesized (Lowry and McMartin, 1974; Schulster *et al.*, 1974), and the proteins were characterized only by one-dimensional gels (Nakamura and Hall, 1978). However, more sensitive methods may well show that these changes occur at early time periods. A more recent study has reported synthesis of 35K protein within a few minutes of exposure to ACTH (Della Cioppa and Hall, 1985). This protein leaves the cytoplasm, passes through the outer membrane and intermembrane fluid to reach the inner membrane (Della Cioppa and Hall, 1985). The fact that these proteins move to mitochondria makes them of interest because it is in this membrane that the response to ACTH is expressed. The time required for the first apparent synthesis of such proteins is difficult to determine.

b. Rat Adrenal. Dazord *et al.* (1978) have reported the synthesis of a cytosolic protein following the administration of ACTH. The protein was first detected after 120 min. A thorough study by Krueger and Orme-Johnson revealed the synthesis of a protein (28K) within 3 min of addition of ACTH to rat adrenal cells (Krueger and Orme-Johnson, 1983). The protein appears at about the same time as increased production of corticosterone—bearing in mind the problems discussed above. This protein appears to result from cotranslational modification of a precursor protein (Krueger and Orme-Johnson, 1983).

In some ways the most intriguing protein so far reported from rat adrenal is the 2.2K cytosolic protein found by Pedersen and Brownie (1983). The interest in this protein lies first in the fact that the protein stimulates side-chain cleavage in mitochondria from rat adrenal, and second because of its small size it could have been overlooked in the usual systems of SDS gels. The mitochondria are prepared from hypophysectomized rats treated with cycloheximide and ACTH to load the outer mitochondrial membrane with cholesterol (Pedersen and Brownie, 1983). The design of these experiments reflects the initial prejudice of the authors concerning the mechanism of action of the protein, namely that it promotes the loading of P-450 with cholesterol (Pedersen and Brownie, 1983). The protein does not apparently bind cholesterol, but the mitochondria treated with the protein show a high proportion of P-450 in the high spin form. With anaerobic mitochondria this could indicate accumulation of ES. The final determination of the amino acid sequence of 2.2K will help to clarify its function. In this connection it is worth noting that the 2.2K protein accelerates side-chain cleavage in mitochondria (Pedersen and Brownie, 1983).

However, 2.2K is not a sterol carrier and the synthesis of sterol carrier protein has not been shown to be increased by ACTH (Pedersen, 1984).

c. *Miscellaneous Proteins.* A number of workers have mixed subcellular fractions of adrenal cells and reported stimulation and inhibition of steroid production (Farese, 1967; Ray and Strott, 1981; Bakker *et al.,* 1978; Neher *et al.,* 1982; Werne *et al.,* 1983). Such studies are preliminary, and since the fractions in question are not purified the resulting effects may represent the algebraic sum of the actions of a variety of substances which make the results difficult to interpret. The inhibitory effect of fractions isolated by Werne *et al.* (1983) is of interest as a reminder of the possible complexity of the system. Inhibition is a less popular aspect of regulation than stimulation, but these findings remind us that inhibition may be important.

4. *The Plasma Membrane*

The plasma membrane makes a number of important contributions to the responses to ACTH, namely, it presents the ACTH receptor for binding of the hormone, it contains the responding adenylate cyclase, and as discussed above, it possesses the LDL receptor that is under control by ACTH. In addition to adenylate cyclase, the plasma membrane contains a tightly bound cyclic AMP-dependent protein kinase that is capable of phosphorylating three intrinsic membrane proteins (Widmaier *et al.,* 1985). These proteins are also phosphorylated in the intact cell under the influence of cAMP, although the functional significance of these changes remains obscure. The plasma membrane participates in receptor-mediated endocytosis of LDL, and in the Y-1 cell, ACTH and cyclic AMP cause the appearance of microvilli on the surface of the cell (Setoguti and Inone, 1981). Although the function of the microvilli are unknown, they may be involved in transport of various substances (e.g., LDL) into the cell (Goshima *et al.,* 1984). Finally, the plasma membrane contains K^+-dependent Ca^{2+} channels that may be important in the regulation of intracellular Ca^{2+} (Kenyon *et al.,* 1985). The plasma membrane also contains cyclic AMP-binding proteins (Wen *et al.,* 1985), the functions of which are not known.

5. *The Nucleus*

The acute steroidogenic response to ACTH is not inhibited by actinomycin D (Ferguson, 1963; Garren *et al.,* 1965) which suggests that the new proteins involved in the response are made on existing mRNA. It should be mentioned here that ACTH and cyclic AMP inhibit cell division by Y-1 cells (Ramachandran and Suyama, 1975). Such a response would be consistent with the widely accepted principle that under certain conditions cells choose between division and differentiation. According to this view, ACTH promotes differentiation at the cost of division.

XII. Synthesis and Conclusions

From these indigestible and confusing facts, what can we say about the steroidogenic action of ACTH?

ACTH binds to a specific receptor in the plasma membrane of its target cell, and this binding immediately activates adenylate cyclase which produces cyclic AMP—the major, if not the only, second messenger. Cyclic AMP immediately promotes phosphorylation of proteins by way of protein kinase A. Membrane proteins are phosphorylated and, no doubt, these changes lead to altered membrane function including, possibly, increase in endocytosis and formation of microvilli. Cholesterol ester hydrolase is activated by phosphorylation, and the active protein releases free cholesterol from depots of cholesterol ester; the free sterol binds to SCP_2. Proteins associated with the cytoskeleton are subjected to phosphorylation and the cytoskeleton reorganizes compartments and surfaces within the cell. One consequence of the changes in the cytoskeleton is increased delivery of cholesterol to the outer mitochondrial membrane.

To this point, newly synthesized proteins are not required—all of these changes take place in the presence of cycloheximide. Now cholesterol must cross the intermembrane space to the inner membrane. This crossing requires new protein(s) and is stimulated by SCP_2. If cholesterol makes the crossing bound to SCP_2, the new protein(s) must assist the protein-cholesterol complex in some way that we cannot now conceive. The new protein(s) need not assist the act of crossing per se; it might, for example, facilitate entry into the inner membrane. There is, at present, no way of knowing. Cholesterol can move slowly from one membrane to another by way of an intervening water phase. However, the rapid transit required in adrenal mitochondria must be regulated—presumably by one or more proteins. It is interesting that 35K protein is rapidly synthesized under the influence of ACTH and moves from cytoplasm to outer membrane and through the intermembrane space to the inner membrane. Presumably, such a protein carries the information within its structure that enables it to proceed to the inner membrane. Evidently, some such protein causes cholesterol to make this crossing. As more information about such processing of intramitochondrial proteins becomes available, we will know what sort of protein to look for.

Even when cholesterol reaches the inner membrane, our conceptual difficulties are not over. We must know whether the cholesterol goes to a specific region or regions of the membrane. When the side-chain cleavage system is reconstituted into vesicles of pure phospholipids and cholesterol, it seems that all the cholesterol is available as substrate for the enzyme (Lambeth and Stevens, 1984). On the other hand, in the mitochondrion it appears that not all the cholesterol of the inner membrane is available to the side-chain cleavage enzyme (Cheng et al., 1985). We know too little about the inner membrane to account for this difference.

When the side-chain cleavage enzyme is finally loaded with cholesterol, pregnenolone is formed and the rest of the pathway to glucocorticoids proceeds rapidly. It is significant that the production of pregnenolone begins with a rapid burst followed by a slower rate of production (Koritz and Moustafa, 1976). This break in the time course of C_{27} side-chain cleavage provides an important clue. Perhaps the enzyme must be reloaded with substrate and again the mechanism of delivery of cholesterol to the enzyme must be explained because the pure enzyme does not show a similar burst followed by slowing of pregnenolone synthesis (Shikita and Hall, 1973a).

Apart from these numerous unsolved problems in the events just described, there looms the question of Ca^{2+}. We still cannot say whether any change in intracellular Ca^{2+} takes place as part of the response to ACTH, and if it does we must learn which step(s) in the response is (are) affected.

In the early days of the studies reviewed above, some workers believed that there might be a single action of cyclic AMP which served, like the lighting of a fuse, to trigger the complete response. It is now clear that cyclic AMP does many things to the cell. It is important that we are not misled simply by the repertoire of available inhibitors. The fact that we cannot specifically inhibit phosphorylation of proteins, for example, does not mean that these changes are not as essential as those we can inhibit, e.g., protein synthesis. Moreover, by considering responses that are not inhibited by cycloheximide, as well as those that are, we can free this inhibitor from the unreasonable responsibility it has been given in the past.

In reflecting upon this outline of the action of ACTH, we might end by considering the future. Immunoelectron microscopy provides the way to finer details of cell structure and the distribution of individual proteins within the cell. Monoclonal antibodies provide high specificity and offer a method for purifying proteins present in small amounts. Videointensification of Nomarski and fluorescent images of living cells enable us to study movement of minute structures. In addition, molecular biology provides details of molecular structure. With these powerful methods we can learn more about the movement of cholesterol through the adrenal cell. This, in turn, will bring us closer to the elusive mechanism of hormone action.

Acknowledgments

The author is extremely grateful to Dr. Eric P. Widmaier for helpful discussions during the preparation of the manuscript and for a critical reading of the final version. The preparation of this manuscript was supported by grants from the National Institutes of Health (AM 28113, AM32236, CA29497).

References

Anderson, J. M., and Dietschy, J. M. (1978). *J. Biol. Chem.* **253,** 9024–9032.
Bakker, C. P., van der Plank-van Wissen, M. P. I. and van der Molen, H. J. (1978). *Biochim. Biophys. Acta* **543,** 235–241.
Bakker, C. P., van der Plank-van Winsen, M. P. I. and van der Molen, H. J. (1979). *Biochim. Biophys. Acta* **584,** 94–103.
Beckett, G. J., and Boyd, G. S. (1977). *Eur. J. Biochem.* **72,** 223–233.
Bell, J. J., and Harding, B. (1974). *Biochim. Biophys. Acta* **348,** 285–290.
Berridge, M. J., and Irvine, R. F. (1984). *Nature (London)* **312,** 315–321.
Berridge, M. J., Dawson, R. M. C., Downes, C. P., Heslop, J. P., and Irvine, R. F. (1983). *Biochem. J.* **212,** 473–482.
Bhargawa, G., Schwartz, E., and Koritz, S. B. (1978). *Proc. Soc. Exp. Biol. Med.* **158,** 183–186.
Birmingham, M. K., Elliot, F. H., and Valere, P. H. L. (1953). *Endocrinology* **53,** 687–695.
Blickstad, I., Markey, F., Carlsson, L., Perrson, T., and Lindberg, U. (1978). *Cell* **15,** 935–943.
Brown, M. S., and Goldstein, J. L. (1976). *Science* **191,** 150–154.
Buckley, D. I., and Ramachandran, J. (1981). *Proc. Natl. Acad. Sci. U.S.A.* **78,** 7431–7435.
Bush, I. W. (1962). *Pharmacol. Rev.* **14,** 317–445.
Cam, G. R., and Bassett, J. R. (1983). *J. Endocrinol.* **98,** 173–182.
Caron, M. G., Goldstein, S., Savard, K., and Marsh, J. M. (1975). *J. Biol. Chem.* **250,** 5137–5143.
Carr, B. R., and Simpson, E. R. (1981). *Endocr. Rev.* **2,** 306–317.
Chanderbhan, R., Noland, B. J., Scallen, T. J., and Vahouny, G. V. (1982). *J. Biol. Chem.* **257,** 8928–8934.
Chang, K., Marcus, N. A., and Cuatrecasas, P. (1974). *J. Biol. Chem.* **249,** 6854–6865.
Charest, R., Pipic, V., Exton, J. H., and Blackmore, P. F. (1985). *Biochem. J.* **227,** 79–90.
Cheitlin, R., Buckley, D. I., and Ramachandran, J. (1985). *J. Biol. Chem.* **260,** 5323–5327.
Cheng, B., Hsu, D. K., and Kimura, T. (1985). *Mol. Cell. Endocrinol.* **40,** 233–240.
Cheung, W. Y. (1980). *Science* **207,** 19–26.
Clegg, J. S. (1984). *Am. J. Physiol.* **246,** R133–R151.
Connely, O. M., Headon, D. R., Olson, C. D., Ungar, F., and Dempsy, M. E. (1984). *Proc. Natl. Acad. Sci. U.S.A.* **81,** 2970–2974.
Crivello, J. F., and Jefcoate, C. R. (1978). *Biochim. Biophys. Acta* **542,** 315–319.

Culty, M., Vilgrain, I., and Chambaz, E. M. (1984). *Biochem. Biophys. Res. Commun.* **129,** 499–503.
Davis, J. S., Farese, R. V., and Clark, M. R. (1983). *Endocrinology* **112,** 2212–2214.
Dazord, A., Gallet, D., and Saez, J. M. (1978). *Biochem. J.* **176,** 233–238.
Della Cioppa, G., and Hall, P. F. (1985). In Preparation.
Downes, G. P., and Michell, R. H. (1981). *Biochem. J.* **198,** 133–140.
Eichorn, J., and Hechter, O. (1957). *Proc. Soc. Exp. Biol. Med.* **95,** 311–314.
Farese, R. V. (1967). *Biochemistry* **6,** 2052–2057.
Farese, R. V. (1983). *Metabolism* **32,** 628–641.
Farese, R. V. (1984). *Mol. Cell. Endocrinol.* **35,** 1–14.
Farese, R. V., Sabir, M. A., and Vandor, S. L. (1979). *J. Biol. Chem.* **254,** 6842–6844.
Farese, R. V., Sabir, M. A., and Larson, R. E. (1980). *J. Biol. Chem.* **255,** 7232–7237.
Farese, R. V., Larson, R. E., and Davis, J. S. (1984). *Endocrinology* **114,** 302–304.
Faust, J. R., Goldstein, J. L., and Brown, M. S. (1977). *J. Biol. Chem.* **252,** 4861–4867.
Ferguson, J. J. (1963). *J. Biol. Chem.* **238,** 2754–2759.
Garren, L. D., Ney, R. L., and Davis, W. W. (1965). *Proc. Natl. Acad. Sci. U.S.A.* **53,** 1443–1447.
Goldstein, J. L., and Brown, M. S. (1974). *J. Biol. Chem.* **249,** 5153–5162.
Goldstein, J. L., Baser, S. K., Brunschede, G. Y., and Brown, M. S. (1976). *Cell* **7,** 85–95.
Goshima, K., Masuda, A., and Owaribe, K. (1984). *J. Cell Biol.* **98,** 801–809.
Gwynne, J. T., and Hess, B. (1980). *J. Biol. Chem.* **255,** 10875–10882.
Gwynne, J. T., and Strauss, J. F., III (1982). *Endocr. Rev.* **3,** 299–317.
Gwynne, J. T., Mahaffee, D., Brewer, H. D., and Ney, R. L. (1976). *Proc. Natl. Acad. Sci. U.S.A.* **73,** 4329–4333.
Hall, P. F. (1967). *Biochemistry* **6,** 2974–2802.
Hall, P. F. (1972). *Biochemistry* **11,** 2891–2897.
Hall, P. F. (1984a). *Can. J. Biochem. Cell Biol.* **62,** 653–665.
Hall, P. F. (1984b). *Int. Rev. Cytol.* **86,** 53–95.
Hall, P. F. (1985). *Recent Prog. Horm. Res.* **41,** 1–39.
Hall, P. F. (1986). *Vitam. Horm.* **42,** in press.
Hall, P. F., Charponnier, C., Nakamura, M., and Gabbiani, G. (1979). *J. Biol. Chem.* **254,** 18:9080–9084.
Hall, P. F., and Nakamura, M. (1979). *J. Biol. Chem.* **254,** 12547–12554.
Hall, P. F., and Young, D. G. (1968). *Endocrinology* **82,** 559–568.
Hall, P. F., Sozer, C. C., and Eik-Nes, K. B. (1964). *Endocrinology* **74,** 35–43, 1964.
Hall, P. F., Lewes, J. L., and Lipson, E. D. (1975). *J. Biol. Chem.* **250,** 6:2283–2286.
Hall, P. F., Nakamura, M., and Mrotek, J. J. (1981a). *Biochim. Biophys. Acta* **676,** 338–344.
Hall, P. F., Osawa, S., and Thomason, C. L. (1981b). *J. Cell Biol.* **90,** 402–407.
Haynes, R. C. (1958). *J. Biol. Chem.* **233,** 1220–1222.
Haynes, R. C., Koritz, S. B., and Peron, F. G. (1959). *J. Biol. Chem.* **234,** 1421–1425.
Hirata, K., Hirota, T., Aguilera, G., and Catt, K. J. (1985). *J. Biol. Chem.* **260,** 3243–3246.
Hsu, D., and Kimura, T. (1982). *Biochem. Biophys. Res. Commun.* **107,** 389–393.
Igarashi, Y., and Kimura, T. (1984). *J. Biol. Chem.* **259,** 10745–10753.
Iida, S., Widmaier, E. P., and Hall, P. F. (1985). Submitted.
Jefcoate, C. R., and Orme-Johnson, W. H. (1975). *J. Biol. Chem.* **250,** 4671–4676.
Kaminami, S., Ochi, H., Kobayashi, Y., and Takemori, S. (1980). *J. Biol. Chem.* **255,** 3386–3394.
Karaboyas, G. C., and Koritz, S. B. (1965). *Biochemistry* **4,** 462–468.
Kenyon, C. J., Young, J., and Fraser, R. (1985). *Endocrinology* **116,** 2279–2285.

Kikkawa, U., Takai, Y., Minakuchi, R., Inohara, S., and Nishizuka, Y. (1982). *J. Biol. Chem.* **257,** 13,341–13,348.
Koritz, S. B., and Moustafa, A. M. (1976). *Arch. Biochem. Biophys.* **174,** 20–26.
Korn, E. D. (1978). *Proc. Natl. Acad. Sci. U.S.A.* **75,** 588–599.
Koroscil, T. M., and Gallant, S. (1980). *J. Biol. Chem.* **255,** 6276–6282.
Kovanen, P. T., Faust, J. R., Brown, M. S., and Goldstein, J. L. (1979). *Endocrinology* **104,** 599–605.
Kraft, A. S., and Anderson, W. B. (1983). *Nature (London)* **301,** 621–623.
Krueger, R. J., and Orme-Johnson, N. R. (1983). *J. Biol. Chem.* **258,** 10159–10167.
Kuo, J. F., and Greengard, P. (1969). *Proc. Natl. Acad. Sci. U.S.A.* **64,** 1349–1353.
Lambeth, J. D., and Stevens, V. L. (1984). *Endocr. Res.* **10,** 283–309.
Lambeth, J. D., Seybert, D. W., Lancaster, J. R., Salerno, J. C., and Kamin, H. (1982). *Mol. Cell. Biochem.* **45,** 13–27.
Lazarides, E. (1980). *Nature (London)* **283,** 249–256.
Long, C. N. H. (1945). *Recent Prog. Horm. Res.* **1,** 99–124.
Lowry, P. J., and McMartin, C. (1974). *Biochem. J.* **142,** 287–291.
Mason, J. I., Arthur, J. R., and Boyd, G. S. (1978). *Biochem. J.* **174,** 1045–1052.
Mathew, J. K., Curtis, J. C., and Mrotek, J. J. (1985). *Steroids* **44,** 105–122.
Mathews, E. K., and Saffran, M. (1973). *J. Physiol. (London)* **234,** 43–64.
Means, A. R., and Dedman, J. R. (1980). *Nature (London)* **285,** 73–78.
Michell, R. H. (1983). *Life Sci.* **32,** 2083–2085.
Mitani, F., Shimizu, T., Ueno, R., Ishimura, Y., Izumi, S., Komatsu, N., and Watanabe, K. (1982). *J. Histochem. Cytochem.* **30,** 1066–1070.
Moore, F. D. (1957). *Recent Prog. Horm. Res.* **13,** 511–536.
Moustafa, A. M., and Koritz, S. B. (1977). *Eur. J. Biochem.* **78,** 231–238.
Mrotek, J. J., and Hall, P. F. (1975). *Biochem. Biophys. Res. Commun.* **64,** 891–896.
Mrotek, J. J., and Hall, P. F. (1977). *Biochemistry* **16,** 3177–3181.
Mrotek, J. J., and Hall, P. F. (1978). *Gen. Pharmacol.* **9,** 269–273.
Naghshinek, S., Treadwell, C. R., Gallo, L. L., and Vahouny, G. V. (1978). *J. Lipid Res.* **19,** 561–569.
Nakajin, S., and Hall, P. F. (1981). *J. Biol. Chem.* **256,** 3871–3876.
Nakajin, S., Shively, J., Yuan, P.-M., and Hall, P. F. (1981). *Biochemistry* **20,** 4037–4042.
Nakajin, S., Shinoda, M., and Hall, P. F. (1983). *Biochem. Biophys. Res. Commun.* **111,** 2:512–517.
Nakajin, S., Shinoda, M., Hanui, M., Shively, J. E., and Hall, P. F. (1984). *J. Biol. Chem.* **259,** 3971–3976.
Nakamura, M., and Hall, P. F. (1978). *Biochim. Biophys. Acta* **542,** 330–339.
Nakamura, M., Watanuki, M., Tilley, B., and Hall, P. F. (1980). *J. Endocrinol.* **84,** 179–188.
Neher, R., Milani, A., Solano, A. R., and Podesta, E. J. (1982). *Proc. Natl. Acad. Sci. U.S.A.* **79,** 1727–1731.
Niedel, J. E., Kuhn, L. J., and Vandenbark, G. R. (1983). *Proc. Natl. Acad. Sci. U.S.A.* **80,** 36–40.
Nishizuka, Y. (1984). *Nature (London)* **308,** 693–698.
Nishizuka, Y., Takai, Y., Tanaka, Y., Miyake, R., and Nishizuka, Y. (1983). *J. Biol. Chem.* **258,** 11442–11445.
Nishizuka, Y., Takai, Y., Kishimoto, A., Kikkawa, Y., and Kaibuchi, K. (1984). *Recent Prog. Horm. Res.* **40,** 301–345.
Ohno, Y., Yanagibashi, K., Yonezawa, Y., Ishiwatari, S., and Matsuba, M. (1983). *Endocrinol. Jpn.* **30,** 335–339.
Olmsted, J. B., and Borisy, G. G. (1973). *Annu. Rev. Biochem.* **42,** 507–540.

Osawa, S., and Hall, P. F. (1984). *Endocrinology,* submitted.
Osawa, S., and Hall, P. F. (1985). *J. Cell Sci.* **32,** 1143–1156.
Osawa, S., Betz, G., and Hall, P. F. (1984). *J. Cell Biol.* **99,** 1335–1342.
Ottaway, J. H., and Mowbray, J. (1977). *Curr. Top. Cell Regul.* **12,** 108–208.
Palmieri, F., and Klingenberg, M. (1979). *In* "Methods in Enzymology" (S. Fleincher and L. Packer, eds.), Vol. 56, pp. 279–283. Academic Press, New York.
Pedersen, R. C. (1984). *Endocr. Res.* **10,** 533–561.
Pedersen, R. C., and Brownie, A. C. (1983). *Proc. Natl. Acad. Sci. U.S.A.* **80,** 1882–1886.
Penning, T. M., and Covey, D. F. (1982). *J. Steroid Biochem.* **16,** 691–699.
Perchellet, J., Shanker, G., and Sharma, R. K. (1978). *Science* **199,** 311–314.
Peron, F., and Koritz, S. B. (1960). *J. Biol. Chem.* **235,** 1625–1632.
Podesta, E. J., Milani, A., Steffen, H., and Neher, R. (1979). *Proc. Natl. Acad. Sci. U.S.A.* **76,** 5187–5191.
Privalle, C. T., Crivello, J. F., and Jefcoate, C. R. (1983). *Proc. Natl. Acad. Sci. U.S.A.* **80,** 702–706.
Rainey, W. E., Shay, J. W., and Mason, J. I. (1984). *Mol. Cell. Endocrinol.* **35,** 189–197.
Ramachandran, J. (1985). *Endocr. Res.* **10,** 347–363.
Ramachandran, J., and Suyama, A. T. (1975). *Proc. Natl. Acad. Sci. U.S.A.* **72,** 113–117.
Ray, P., and Strott, C. A. (1981). *Life Sci.* **28,** 1529–1533.
Remmer, H., Schenkman, J., Estabrook, R. W., Sasame, H., Gillette, J., Harasimhulu, S., Cooper, D. Y., and Rosenthal, O. (1966). *Mol. Pharmacol.* **2,** 187–193.
Ris, H. (1985). *J. Cell Biol.* **100,** 1474–1487.
Sala, G. B., Hayashi, K., Catt, K. J., and Dufau, M. L. (1979). *J. Biol. Chem.* **254,** 3861–3865.
Samuels, L. T. (1960). *Metab. Pathways* **1,** 431–480.
Schlegel, R. A., and Reichsteiner, M. C. (1975). *Cell* **5,** 371–374.
Schimmer, B. P. (1980). *Adv. Cyclic Nucleotide Res.* **13,** 181–214.
Schimmer, B. P., and Zimmerman, A. E. (1976). *Mol. Cell. Endocrinol.* **4,** 263–270.
Schulster, D., Richardson, M. C., and Palfreyman, J. W. (1974). *Mol. Cell. Endocrinol.* **2,** 17–21.
Selye, H. (1954). *J. Clin. Endocrinol.* **14,** 997–1005.
Setoguti, T., and Inone, Y. (1981). *Acta Anat.* **111,** 207–221.
Shikita, M., and Hall, P. F. (1973a). *J. Biol. Chem.* **248,** 5598–5604.
Shikita, M., and Hall, P. F. (1973b). *J. Biol. Chem.* **248,** 5605–5609.
Shikita, M., and Hall, P. F. (1974). *Proc. Natl. Acad. Sci. U.S.A.* **71,** 4, 1441–1445.
Stone, D., and Hechter, O. (1954). *Arch. Biochem. Biophys.* **51,** 457–461.
Taylor, D. L., and Condeelis, J. S. (1979). *Int. Rev. Cytol.* **56,** 57–144.
Thomas, A. P., Alexander, J., and Williamson, J. R. (1984). *J. Biol. Chem.* **259,** 5574–5584.
Trzeciak, W. H., and Boyd, G. S. (1974). *Eur. J. Biochem.* **46,** 201–207.
Vahouny, G. V., Chanderbhan, R., Noland, B. J., Irwin, D., Dennis, P., Lambeth, J. D., and Scallen, T. J. (1983). *J. Biol. Chem.* **258,** 11731–11737.
Vahouny, G. V., Chanderbhan, R., Noland, B. J., and Scallen, T. J. (1985). *Endocr. Res.* **10,** 473–505.
Vaughan, M. H., Jr., Pawlowski, P. J., and Forehhammer, J. (1971). *Proc. Natl. Acad. Sci. U.S.A.* **68,** 2057–2061.
Vershoor-Klootwyk, A. H., Vershoor, L., Azhar, S., and Reaven, G. M. (1982). *J. Biol. Chem.* **257,** 7666–7671.
Vilgrain, I., Defaye, G., and Chambaz, E. M. (1984). *Biochem. Biophys. Res. Commun.* **125,** 554–561.

Vulliet, P. R., Woodgett, J. R., Ferrari, S., and Hardie, D. G. (1985). *FEBS Lett.* **182,** 335–339.
Warren, L., and Glick, M. C. (1969). In "Fundamental Techniques in Virology," pp. 66–71. Academic Press, New York.
Warren, L., Glick, M. C., and Nass, M. K. (1966). *J. Cell. Physiol.* **68,** 269–287.
Watanuki, M., Tilley, B. E., and Hall, P. F. (1977). *Biochim. Biophys. Acta* **483,** 236–247.
Watanuki, M., Tilley, B. E., and Hall, P. F. (1978). *Biochemistry* **17,** 127–130.
Wen, S. C., Chang., C., Reitherman, R. W., and Harding, B. W. (1985). *Endocrinology* **116,** 935–944.
Werne, P. A., Greenfield, N. J., and Lieberman, S. (1983). *Proc. Natl. Acad. Sci. U.S.A.* **80,** 1877–1881.
Widmaier, E. P., and Hall, P. F. (1985). Mol. Cell *Endocrinology* **43,** 181–188.
Widmaier, E. P., Osawa, S., and Hall. P. F. (1985). *Endocrinology* **118,** 701–708.
Williamson, J. R., Cooper, R. H., Joseph, S. K., and Thomas, A. P. (1985). *Am. J. Physiol.* **248,** C203–C216.
Wolosewick, J. J., and Porter, K. R. (1979). *J. Cell. Biol.* **82,** 114–139.
Yago, N., Kobayashi, S., Kekeyama, S., Jurokawa, H., Iwai, Y., Suzuki, T., and Ichii, S. (1970). *J. Biochem.* **68,** 775–783.
Yanagibashi, K. (1979). *Endocrinol. Jpn.* **26,** 227–232.
Yanagibashi, K., Kamuja, N., Lin, G., and Matsuba, M. (1978). *Endocrinol. Jpn.* **25,** 545–551.
Yasamura, Y. (1968). *Am. Zool.* **8,** 285–288.
Yuan, P. M., Nakajin, S., Hanniu, M., Hall, P. F., and Shively, J. E. (1983). *Biochemistry* **22,** 143–149.

5

Effect of ACTH and Other Proopiomelanocortin-Derived Peptides on Aldosterone Secretion

ALEXANDER C. BROWNIE
AND ROBERT C. PEDERSEN

Department of Biochemistry
Schools of Medicine and Dentistry
State University of New York at Buffalo
Buffalo, New York 14214

I. Introduction

Much of the active research on the control of aldosterone biosynthesis has focused on the roles of sodium, potassium, and the renin–angiotensin system. However, soon after the isolation and characterization of aldosterone it became evident that the adenohypophysis also exerts an influence on aldosterone secretion (Davis, 1976). Studies with hypophysectomized laboratory animals and of human panhypopituitarism have been especially revealing with respect to potential relationships between pituitary hormones and aldosterone. They establish that in the absence of proper pituitary function there is an acute fall in aldosterone secretion and a diminished capacity for adrenocortical response to sodium depletion.

This effect of hypophysectomy can be explained in part by postulating a role for adrenocorticotropic hormone (ACTH), and, indeed, it is well-established that ACTH treatment acutely stimulates aldosterone secretion. However, there is also strong evidence that pituitary hormones other than ACTH, including growth hormone and prolactin, are involved in this control (Palmore and Mulrow, 1967; McCaa *et al.*, 1974). Also, Sen *et al.* (1977, 1981a,b) have reported that aldosterone stimulating factor, a pituitary glycoprotein isolated from urine, regulates aldosterone secretion. More recently, the demonstration that ACTH is derived from a larger polyprotein, proopiomelanocortin (POMC), from which other biologically active segments are also processed, has refocused attention on ACTH-related peptides. Several groups have examined the potential role for these POMC-derived peptides in the control of corticosteroidogenesis. In this review we will address the mechanism of action of ACTH on aldosterone biosynthesis and, in addition, the evidence for and against roles for β-lipotropin (β-LPH), β-endorphin, β-melanotropin (β-MSH), α-MSH, and pro-γ-MSHs in the control of that pathway.

II. Aldosterone Biosynthetic Pathway

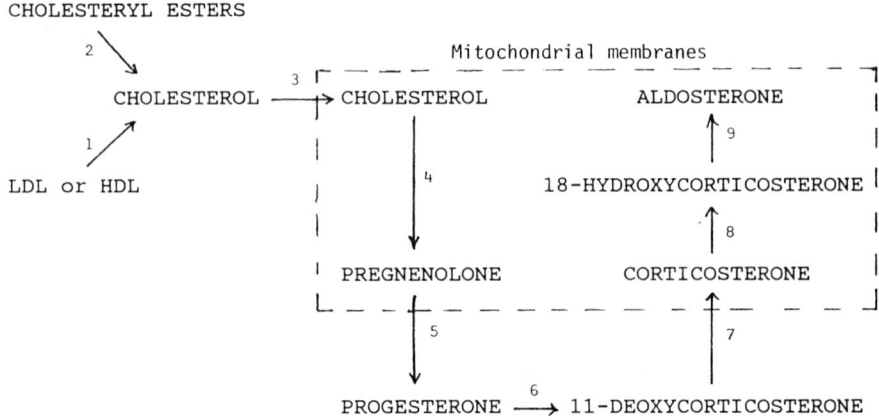

The pathway for aldosterone biosynthesis is summarized above. Reaction **1** involves binding of low-density lipoprotein (LDL) or high-density lipoprotein (HDL) particles to specific receptors on the plasma membrane of the zona glomerulosa (ZG) cell and lipoprotein processing within the cell to yield free cholesterol. Reaction **2**, catalyzed by a cytosolic cholesteryl ester hydrolase, supplies free cholesterol for steroidogenesis. Reaction **3** is the transport of cholesterol to the inner mitochondrial membrane, facilitated by a sterol carrier protein [cf. SCP$_2$ in the zona fasciculata (ZF)]. Reaction **4**, the regulated step in what is usually denoted as the "early pathway," is catalyzed by the side-chain cleavage cytochrome *P*-450 complex in the inner mitochondrial membrane. This reaction, with 20,22-dihydroxycholesterol as an intermediate, requires O_2 and NADPH. In reaction **5**, pregnenolone leaves the mitochondrion for oxidation to progesterone by NAD^+-linked 3β-hydroxysteroid dehydrogenase/isomerase on the smooth endoplasmic reticulum (SER). Reaction 6 is catalyzed by cytochrome P-450_{21}, also on the SER. Reaction 7 is the translocation of 11-deoxycorticosterone to the mitochondrion and its 11β-hydroxylation there, catalyzed by cytochrome P-$450_{11\beta}$ in the inner membrane.

Various details of the "late pathway"—reactions 8 and 9—have been elucidated more recently. This pathway, from corticosterone to aldosterone by way of 18-hydroxycorticosterone, is frequently depicted as a dehydrogenation reaction. However, there is now a wealth of evidence that the final reactions in aldosterone biosynthesis are mixed-function oxidations. Ulick (1972, 1976) first suggested this based on his studies of "late pathway" dysfunction. Marusic *et al.* (1973) demonstrated that the conversion of corticosterone and 18-hydroxycorticosterone to aldosterone required O_2 and NADPH, the latter generated from oxidation of TCA cycle intermediates (Aupetit *et al.*, 1983). Moreover, because the conversion of 18-hydroxycorticosterone to aldosterone is blocked by CO and other inhibitors of cytochrome P-450s (Kojima *et al.*, 1982, 1984a), the enzyme or enzymes involved are likely members of that family.

From their studies of 18-hydroxycorticosterone metabolism in salt-sensitive and -resistant strains of rats, Rapp and Dahl (1976) proposed two successive hydroxylations, the first catalyzed by a "corticosterone methyloxidase I" (CMO I) activity to generate 18-hydroxycorticosterone, and the second catalyzed by a "corticosterone methyloxidase II" (CMO II) to form aldosterone. More recent work by Hall and colleagues (Yanagibashi *et al.*, 1986) with bovine adrenal cortex suggests that 11β-/18-hydroxylase is the sole enzyme required for these "late pathway" reactions. Although the precise locus of hormonal control in the late pathway is still unresolved, it is probably not the CMO II step because, with certain exceptions (e.g., CMO II deficiency or the 17α-hydroxylase form of congenital adrenal hyperplasia), the ratio of plasma 18-hydroxycorticosterone to aldosterone remains remarkably constant (Kater *et al.*, 1985).

III. ACTH Action on Aldosterone Biosynthesis

A. ACTH Receptors on Zona Glomerulosa Cells

Buckley *et al.* (1981) were first to synthesize an ACTH homolog whose radioiodinated adduct retains full adrenotropic potency. These investigators (Buckley and Ramachandran, 1981) then used the peptide [Phe^2, Nle^4]ACTH$_{1-24}$, to characterize ACTH receptors associated with cells isolated from the zona fasciculata/reticularis (ZF/R). This homolog has recently been employed for the same purpose with rat ZG cells (Gallo-

Payet and Escher, 1985). Both high- and low-affinity binding sites were observed. Surprisingly, the ACTH receptor density on ZG cells appears to be much greater than on ZF/R cells. The physiological significance of this is unclear.

B. ROLE OF cAMP, Ca^{2+}, EICOSANOIDS, AND ANF IN ACTH ACTION ON ALDOSTERONE BIOSYNTHESIS

As with the ZF, ACTH stimulation of ZG steroidogenesis is dependent upon cAMP as a second messenger. This has been established in a number of ways. Kaplan (1965) showed that treatment of tissue from the ZG with cAMP results in increased aldosterone secretion. Other investigators have demonstrated a positive effect of ACTH on cAMP levels using outer slices of beef adrenals (Saruta et al., 1972) and isolated ZG cells (Albano et al., 1974; Fujita et al., 1979). Furthermore, the ACTH-stimulated activation of adenylate cyclase in ZG cells correlates with cAMP accumulation and aldosterone secretion (Douglas et al., 1978).

In contrast, it is now well established (Saruta et al., 1972; Peytremann et al., 1974; Douglas et al., 1978; Bell et al., 1981) that control of aldosterone biosynthesis by angiotensin II depends not on cAMP but rather on changes in the intracellular concentration of Ca^{2+}. Nevertheless, despite a clear distinction between ACTH and angiotensin II in mechanism of action, there are some important interrelationships. For example, chronic treatment with ACTH blunts the ZG response to angiotensin II, probably by promoting a down-regulation of the angiotensin II receptor (Andoka et al., 1984).

Moreover, Ca^{2+} is also important for ACTH action; suboptimal concentrations of extracellular Ca^{2+} (Fakunding et al., 1979) or Ca^{2+} channel blockers (Fakunding and Catt, 1980) impair the response of ZG cells to the hormone, perhaps by attenuating adenylate cyclase activity. Also, inhibitors of calmodulin adversely affect ACTH-stimulated aldosterone secretion (Balla et al., 1982; Wilson et al., 1984). Recently, the role of Ca^{2+} in aldosteronogenesis has received attention in studies carried out by Rasmussen and colleagues. They report that ACTH, K^+, and angiotensin II each stimulate a Ca^{2+} influx across the plasma membrane (Kojima et al., 1985a,b), though only angiotensin II fosters a rapid, transient mobilization of Ca^{2+} from stores within the cell, apparently by elevating the cytosolic content of triphosphoinositol (Kojima et al., 1984b). Spät and co-workers (Balla et al., 1984; Enyedi et al., 1985) have confirmed the failure of ACTH to activate phosphoinositide turnover in rat ZG cells. However, Braley et al. (1985), using the Ca^{2+}-sensitive indicator quin2, report that they are unable to detect *any* cAMP- or ACTH-dependent change in the intracellular Ca^{2+} concentration of the ZG.

A number of substances has been proposed as modulators of ACTH action on the ZG. Among these are the eicosanoids, of special interest because arachidonate, their precursor, is a major ester of cholesterol in the adrenal cortex. The zone does in fact synthesize prostaglandin E_2 (PGE_2) and prostaglandin I_2 (PGI_2) (Campbell and Gomez-Sanchez, 1985), but the amounts are low and their physiological significance in this regard is questionable (Miller *et al.*, 1980). For example, prostaglandins of the E series potentiate aldosterone production by dispersed ZG cells, but only at very high concentrations (Saruta and Kaplan, 1972; Enyedi *et al.*, 1981).

The atrial natriuretic factors (ANF) comprise another set of potential ACTH modulators under active investigation. These peptides appear to blunt the *in vitro* effectiveness of secretagogues, including ACTH, on bovine ZG cells (Goodfriend, *et al.*, 1984; De Lean *et al.*, 1984; Racz *et al.*, 1985) and rat ZG cells (Atarashi *et al.*, 1984, 1985; Chartier *et al.*, 1984; Kudo and Baird, 1985; Campbell *et al.*, 1985; Chartier and Schiffrin, 1986a). The underlying mechanism for this action is still unclear but the possibilities include an alteration in Ca^{2+} metabolism (Chartier and Schiffrin, 1986b) or a change in the activity of adenylate cyclase (Anand-Srivastava *et al.*, 1985) or the particulate isoform of guanylate cyclase (Waldman *et al.*, 1985; Tremblay *et al.*, 1986). Although data for the late pathway are conflicting, the influence of ANF is clearly manifest on the ACTH-stimulated early pathway. Furthermore, unlike the prostaglandins, ANF appears to act at concentrations that are meaningful. The full significance of these findings for cardiovascular homeostasis are not yet clear.

C. Cholesterol Uptake by Zona Glomerulosa Cells

It is now recognized that substrate cholesterol for adrenal steroidogenesis in the inner zones is derived primarily from plasma lipoproteins (M. S. Brown *et al.*, 1979; Andersen and Dietschy, 1978, 1981; Gwynne and Strauss, 1982). With respect to the ZG, studies by Nagy *et al.* (1984) with cells isolated from rats pretreated with 4-aminopyrazolo[3,4-*d*]pyrimidine (4-APP), a hypocholesterolemic agent, revealed an enhancement of both basal and ACTH-stimulated aldosterone production when incubations were supplemented with lipoproteins. Unfortunately, the behavior of cells derived from untreated controls was not examined, so these data are difficult to assess in context. However, there is evidence that a depletion of rat plasma lipoproteins with 4-APP has little effect on either the concentration of lipid droplets in ZG cells or the aldosteronogenic response to ACTH (Szabo *et al.*, 1984). This is in sharp contrast to the effect of 4-APP treatment on the inner zones. Indeed, given the particularly abundant reserves of cholesteryl esters in the ZG, it is unlikely that those factors which influence aldosterone secretion exert a major part of their tropic action by modulating lipoprotein metabolism.

Taking a more physiological approach, Gross *et al.* (1981) used adrenal scintigraphy to assess the effect of dietary sodium on adrenal uptake of cholesterol in dexamethasone-suppressed dogs. They found that both the cholesterol uptake and aldosterone production varied inversely with the level of salt intake. Furthermore, at least half of the total cholesterol uptake was responsive to ACTH. However, the degree to which this response was specific for the ZG was not assessed.

Finally, although cytosolic cholesteryl ester hydrolase has been investigated rather extensively in the ZF/R as a potential site for steroidogenic control (Pedersen and Brownie, 1986), to our knowledge there are no published data addressing the effect of ACTH on the comparable ZG enzyme.

D. Control of the Early and Late Pathways

Numerous studies indicate that ACTH increases the activity of the early pathway of aldosterone biosynthesis. For example, Kaplan and Bartter (1962) found that when steroid intermediates distal to pregnenolone in the pathway were added to slices enriched with ZG tissue, there was no effect of ACTH on aldosterone output. However, ACTH did increase the rate at which cholesterol was metabolized to aldosterone. Müller (1966) repeated these studies successfully using radiolabeled precursors.

In order to define those reactions under ACTH control, Aguilera and Catt (1979) employed various steroidogenic inhibitors with dispersed ZG cells from dogs and rats. They found that the hormone stimulated both the formation of pregnenolone (i.e., early pathway activity) and, confirming the study by Haning *et al.* (1970), the conversion of corticosterone to aldosterone (late pathway activity). The former reaction is catalyzed by the cytochrome $P\text{-}450_{scc}$ complex in the inner mitochondrial membrane and has been studied extensively in ZF/R tissue. In that venue, ACTH promotes an interaction between cholesterol and the enzyme active site which is facilitated by a polypeptide activator (Pedersen and Brownie, 1983b). A similar mechanism for ACTH action in the ZG has not yet been established, but it is noteworthy that angiotensin II promotes cholesterol–

cytochrome P-450$_{scc}$ association in ZG tissue from sodium-depleted rats (Kramer *et al.*, 1979, 1980).

Aguilera and Catt (1979) have demonstrated that, although the predominant effect of ACTH on steroidogenesis in the ZG is centered on the early pathway, the hormone also increases the rate of conversion of corticosterone to aldosterone. A characteristic type I absorbance change is associated with the binding of corticosterone to cytochrome P-450$_{18}$ in the ZG (Kramer *et al.*, 1979), and, properly exploited, this property could facilitate a more detailed examination of the mechanism by which ACTH modulates the final steps of aldosterone biosynthesis.

IV. Non-ACTH Pituitary Factors Controlling Aldosterone Secretion

The most compelling evidence for non-ACTH pituitary factors that influence aldosterone secretion comes from studies that rely on sodium depletion as a probe. Patients with panhypopituitarism produce normal amounts of aldosterone under innocuous circumstances, but they do not respond to sodium depletion with a normal compensatory increase in aldosterone secretion (Lieberman and Luetscher, 1960; Williams *et al.*, 1971) and the problem is not corrected by replacement therapy with growth hormone (McCaa *et al.*, 1978).

In sodium-depleted rats, Palmore and Mulrow (1967) found that hypophysectomy abolished the customary increase in aldosterone production. Moreover, the administration of ACTH or growth hormone to these animals failed to reestablish the capacity for a full aldosterone response (Palmore and Mulrow, 1967; Lee and de Weid, 1968; Palmore *et al.*, 1970). In studies with dogs, McCaa *et al.* (1974) demonstrated that nephrectomy (i.e., a compromised renin–angiotensin system) did not block the normal aldosterone rise in response to low sodium as long as the pituitary gland was left intact. Aldosterone secretion also was not impaired in nephrectomized dogs suppressed with dexamethasone, further suggesting that the hypothetical pituitary factor is not ACTH.

On the other hand, Palmore and Mulrow (1967) demonstrated that injection of a whole pituitary homogenate fully restored the sodium-dependent aldosterone response in hypophysectomized rats. Lee *et al.* (1968) obtained similar results using incubations of adrenals from hypophysectomized rats; the pretreatment of these animals with anterior pituitary extracts restored the response to sodium depletion as measured by adrenal aldosterone secretion *in vitro*. Moreover, Solyom *et al.* (1971) found that, by adding to adrenal tissue *in vitro* the spent medium from incubations of pituitaries derived from sodium-depleted rats, aldosterone secretion was dramatically enhanced.

More recently, the discovery of POMC in the pituitary corticotroph has directed the attention of investigators to polypeptides processed from it as potential regulators of aldosterone biosynthesis. α-MSH, β-LPH, β-MSH, β-endorphin, and pro-γ-MSHs have all been tested, and the evidence in support of each is considered below.

A. α-MSH

α-MSH, secreted primarily from cells of the intermediate pituitary, has been recognized as a corticotroph product for many years. Its sequence comprises the first 13 amino acids of ACTH, with posttranslational modifications at both the N- and C-termini (acetylation and amidation, respectively). The peptide is a full ACTH agonist with respect to corticosterone production by rat adrenocortical tissue ($ED_{50} \cong 10^{-6}$ M) (Lowry and McMartin, 1972; Seelig and Sayers, 1973), and there is evidence for its substantial influence on the fetal adrenal (Silman *et al.*, 1976; Swaab and Visser, 1977; Llanos *et al.*, 1979; Glickman *et al.*, 1979).

In the adult rat, α-MSH circulates at concentrations between 10^{-11} and 10^{-10} M under basal conditions (Wilson and Morgan, 1979; Thody *et al.*, 1980) and can rise as much as 5-fold above this range with stress (Wilson and Harry, 1980). In humans, data from normal controls and from patients with various forms of pituitary–adrenal dysfunction indicate that plasma α-MSH levels are somewhat lower (Thody *et al.*, 1985).

1. *α-MSH Effects on Aldosterone Secretion*

Over the last decade investigators have examined α-MSH for a potential role in the increased responsiveness of the adrenal cortex during sodium depletion. Page *et al.* (1974) found initially that treatment of sodium-depleted, hypophysectomized rats with growth hormone and α-MSH together restored the normal aldosterone secretory response. Subsequently, Vinson and co-workers carried out a series of studies addressing the effects of α-MSH on aldosterone production by isolated rat ZG cells. They demonstrated first that α-MSH was the active aldosterone-stimulating factor present in posterior pituitary extracts (Vinson *et al.*, 1980). The peptide increased aldosterone secretion from ZG cells at concentrations of 10^{-9} M and higher ($ED_{50} \cong 10^{-8}$ M) but failed to stimulate glucocorticoid production by ZF/R cells, even in very large doses.

Others (Szalay and Stark, 1982; Li *et al.*, 1982; Jornot *et al.*, 1985) have confirmed the aldosteronogenic potency of α-MSH, but Li *et al.* (1982)

found it to be a secretagogue on the inner zones as well. Recently, Vinson's group has reexamined this discrepancy, using perfused rat adrenal glands *in situ* (Hinson *et al.*, 1985) and reports that high concentrations of α-MSH will indeed promote corticosterone secretion. At lower doses of α-MSH, aldosterone secretion was selectively stimulated. Although under sodium-replete conditions, ACTH was three orders of magnitude more potent than α-MSH in this respect, the minimum effective dose for α-MSH fell from 10^{-7} to 10^{-10} M when ZG cells from sodium-depleted rats were substituted (Vinson *et al.*, 1981a,b; 1983). The level of circulating α-MSH is not itself altered by dietary sodium, but this reported increase in adrenal sensitivity to the peptide is sufficient to shift the potency of α-MSH into a range that is potentially physiological.

Shenker *et al.* (1985) observed that the decrement in circulating aldosterone in hypophysectomized, sodium-depleted rats, as compared with intact controls, could be reversed with α-MSH (8 μg/day) but not with ACTH (6 μg/day). However, α-MSH had no influence on the level of plasma aldosterone in intact animals. The explanation for this increased sensitivity of the ZG to α-MSH, both *in vitro* and *in vivo*, following sodium depletion is unknown. The data do suggest, however, that, to the extent α-MSH plays a special role in aldosterone biosynthesis, it may be limited to this setting.

In addition to its potential as an independent agonist, Szalay and Stark (1982) observed that concentrations of α-MSH ranging between 10^{-9} and 6×10^{-7} M potentiate the action of ACTH on aldosterone production by rat ZG cells. Li *et al.* (1982) reported that submaximal α-MSH and ACTH concentrations together produced an additive effect on aldosterone biosynthesis by this cell type. This coordinate action of α-MSH and ACTH is similar to that reported for γ_3-MSH and ACTH (Pedersen *et al.*, 1980), as discussed below (Section IV,C,1), and for α-MSH and γ_3-MSH (Vinson *et al.*, 1984).

2. Mechanism of Action of α-MSH

In attempts to define the mechanism of action for α-MSH, as well as for other secretagogues, investigators have had to be mindful of the potential complications which can arise from the variable but inevitable ZF/R contamination of their preparations. Vinson *et al.* (1983) report no change in cAMP at concentrations of α-MSH just potent enough to elicit a modest aldosteronogenic response. This conflicts with data from Hyatt *et al.* (1985), who do observe a dose-dependent rise. Moreover, Li *et al.* (1982) have noted that α-MSH dose–response curves for aldosterone production by rat ZG cells parallel those for ACTH, suggesting that these peptides may share the same membrane receptor. If true, it would support a role for cAMP as an α-MSH second messenger, since the ACTH receptor is coupled to adenylate cyclase. Finally, Hyatt et al. (1985) fail to discern an α-MSH effect on phosphoinositide metabolism in ZG cells. Though it is not conclusive, this finding is probably inconsistent with a role for intracellular Ca^{2+} as a primary mediator of α-MSH activity.

B. β-LPH, β-MSH, AND β-ENDORPHIN

β-LPH comprises the C-terminal region of POMC, adjacent to and downstream from ACTH. β-LPH itself consists both of β-endorphin, a 31-amino acid sequence at the C-terminus of β-LPH, and of γ-LPH, several segments of which are not well conserved. β-MSH resides in a central portion of β-LPH (β-LPH_{35-56} in man).

There is a voluminous literature, outside the scope of this review, that addresses the levels of circulating β-LPH/γ-LPH and β-endorphin in rats and humans (e.g., Krieger et al., 1977; Tanaka et al., 1978; Nakao et al., 1978; McLoughlin et al., 1980). The range for plasma β-LPH and β-endorphin in normal subjects is 10^{-11}–10^{-10} M, the upper limit of which can be surpassed several-fold when the hypophysis is challenged or dysfunctional. Whereas the authenticity of these circulating polypeptides has been well documented, most of the modern data for β-MSH suggests that, except in cases of ectopic generation, it does not circulate as such (Tanaka et al., 1977). Consequently, investigations of β-MSH potency on the adrenal cortex must be evaluated with an implicit assumption that the peptide is an "active element" of its parent, β-LPH, and can substitute for it (or for γ-LPH) experimentally.

1. *Effects on Aldosterone Secretion*

   a. β-LPH. In an early study, R. D. Brown et al. (1979) failed to detect any effect of β-MSH at concentrations as high as 10^{-6} M on aldosterone production by isolated bovine ZG cells. However, more recently, Matsuoka et al. (1980a; 1980b) observed that treatment of rats with $β_o$-LPH increased aldosterone secretion *in vivo*. They also found that β-LPH from sheep and human pituitaries, as well as synthetic $β_o$-LPH, stimulated aldosterone secretion by isolated ZG cells; small but significant effects were detected with 10^{-9} M β-LPH, and the response was half maximal at 10^{-7} M. Because the purified β-LPHs did not increase corticosterone secretion by ZF/R cells, they probably were not contaminated with ACTH. In contrast, although Washburn et al. (1982) achieved similar results with highly purified $β_h$-LPH and $β_o$-LPH, they did report an immunoreactive (IR)-ACTH contamination of their preparations that was sufficient to account for part of the adrenotropic effect.

b. β-MSH. van der Wal and de Weid (1968) initially reported that β-MSH was ineffective with rat adrenal quarters, but this work preceded the development of sensitive radioimmunoassays for aldosterone. Matsuoka *et al.* (1981a,b) found that β-MSH increased aldosterone secretion by rat ZG cells with an ED_{50} comparable to that for β-LPH. Like its parent, β-MSH did not stimulate corticosterone production by cells from the inner zones. These data have been corroborated in part by Li *et al.* (1982) and by Washburn *et al.* (1982) with synthetic $β_h$-MSH. However, experiments with homologous β-MSH, potentially worthwhile in view of the poor conservation of sequence noted above, have yet to be carried out in the rat. Finally, it is noteworthy that Yamakado *et al.* (1985) have shown an effect of β-MSH on both the early and late aldosterone biosynthetic pathways, both of which also respond to ACTH and angiotensin II.

c. β-Endorphin. The data concerning β-endorphin and aldosteronogenesis are widely conflicting. Matsuoka *et al.* (1981a) found no effect of the neuropeptide on ZG cells over a wide range of concentrations. Similar results were achieved by Washburn *et al.* (1982) with synthetic homologs of β-endorphin. Indeed, Szalay and Stark (1981) reported that, at 10^{-9}–10^{-7} M, β-endorphin actually inhibited the aldosterone response to ACTH. Others have reported a diminished adrenocortical activity in the presence of a related opioid, Met-enkephalin (Rácz *et al.*, 1980, 1982).

On the other hand, experiments by Lymangrover *et al.* (1983) with naloxone, a morphine antagonist, suggest a role for some endogenous opioid in adrenal control. Moreover, Güllner and Gill (1983) observed a significant rise in aldosterone secretion in dogs infused with β-endorphin. Whether this constituted a primary effect on the ZG was not investigated. This is an important consideration in view of reports by Rabinowe *et al.* (1985) that β-endorphin, when administered to normal subjects on normal or low-sodium diets, stimulated an increment in both plasma renin activity and circulating aldosterone, and by Kem *et al.* (1985) that infusions of β-endorphin did *not*, in fact, increase plasma aldosterone in humans. Although these disparate findings may suggest a complex control of aldosterone secretion by β-endorphin, we believe the evidence, on balance, requires that any such putative modulation be indirect.

2. *Mechanism of Action of β-LPH and β-MSH*

The mechanism by which these polypeptides influence aldosterone secretion is unclear. Washburn *et al.* (1982) suggested that the effects of β-LPH and β-MSH were in part a function of the heptapeptide sequence they share with ACTH, i.e., that the stimulation is "ACTH-like." Although this might suggest a role for cAMP in β-LPH/β-MSH activity, Matsuoka et al. (1980b) found no evidence for increased cAMP when ZG cells were stimulated with β-LPH.

3. β-LPH, β-MSH, and β-Endorphin in Hyperaldosteronism

Griffing et al. (1985a) have detected increased concentrations of β-endorphin in 10 patients with idiopathic hyperaldosteronism (IHA), whereas levels were unremarkable in 4 subjects with aldosteronomas. These data contrast with findings of Güllner et al. (1983), who reported no change in total β-endorphin/β-LPH in 2 patients with IHA. One plausible explanation for this discrepancy could be the more selective nature of the assay employed by the former group, thereby unmasking changes specific for plasma β-endorphin.

C. PRO-γ-MSHs

The first demonstration that polypeptides derived from the N-terminal portion of POMC (N-POMC) can influence corticosteroid biosynthesis came from our studies with the murine "16K fragment" (comprising most or all of N-POMC) isolated by Eipper and Mains (1978). Attention was initially drawn to that part of the prohormone when a third region of MSH sequence homology, located within N-POMC, was inferred from the cDNA sequence encrypting bovine POMC (Nakanishi et al., 1979). This MSH sequence was denoted γ-MSH and was subsequently shown to be highly conserved, consistent with some potential biological function. Although γ-MSH itself is probably not a significant product of POMC processing, larger fragments of N-POMC containing γ-MSH are in fact generated, and these are collectively denoted here as pro-γ-MSHs. One of them, γ_3-MSH, is a 27-mer consisting of γ-MSH with a C-terminal extension. It appears that the exact profile of secreted pro-γ-MSHs is dependent on a number of factors, including species, state of the hypothalamic–pituitary–adrenal axis, and the pituitary lobe of origin.

In the rat, we have shown that two heterogeneous IR forms of γ-MSH are present in plasma (Pedersen et al., 1982). The smaller of these probably corresponds to glycosylated γ_3-MSH and the larger to rat 16K fragment. Following stress, plasma IR-γ-MSHs and ACTH rise together—the smaller form of IR-γ-MSH most dramatically. Hale et al. (1984) reported that plasma levels of IR-γ-MSH and ACTH increase concomitantly in human subjects during insulin-induced hypoglycemia or following bolus injections of synthetic corticotropin releasing factor (CRF) and decline together during dexamethasone suppression. This group also established the presence of a circadian rhythm in plasma IR-γ-MSH similar to that for ACTH. The basis for a direct secretory relationship between pro-γ-MSHs and ACTH was confirmed by Chan *et al.* (1982) with their observation that CRF stimulated the simultaneous release of ACTH and N-POMC products from cultured human pituitary cells. These studies, as well as reports by Chan *et al.* (1983) and Motomatsu *et al.* (1984), provide support for the concept of pro-γ-MSHs as authentic hormones.

1. *Pro-γ-MSH Effects on Aldosterone Secretion*

We showed initially that mild trypsin treatment of 16K fragment generated a product (or products) which synergized with ACTH in stimulating corticosterone production by rat ZF/R cells (Pedersen and Brownie, 1980). Soon thereafter, the availability of synthetic peptides containing the γ-MSH sequence (Ling *et al.*, 1979) facilitated experiments (Pedersen *et al.*, 1980) which established that γ$_3$-MSH had the activity originally found in trypsinized 16K fragment. The administration of γ$_3$-MSH to hypophysectomied rats led to greatly augmented plasma corticosterone levels in response to ACTH. Unexpectedly, there was a similar effect on the aldosterone response to ACTH. γ$_3$-MSH had virtually no capacity to increase corticosterone or aldosterone secretion in the absence of ACTH, suggesting that these peptides work in coordinate fashion to control corticosteroidogenesis. In experiments with rat ZF/R cells, the potency of γ$_3$-MSH proved to be substantial; the ED$_{50}$ for potentiation of 10^{-12} M ACTH by rat γ$_3$-MSH was 1.7×10^{-12} M (Pedersen and Brownie, 1983a).

In related work with perfusions of rat ZG and ZF/R cells, Al-Dujaili *et al.* (1981) demonstrated that human 16K fragment, N-POMC$_{1-76}$, also potentiates the ACTH response. No significant increase in corticosterone or aldosterone secretion occurred with N-POMC$_{1-76}$ alone. In comparable perfusions of a mixture of human ZG and ZF/R cells, corticosterone secretion was doubled and aldosterone secretion rose 20–40% with N-POMC$_{1-76}$ plus ACTH as compared with ACTH only. The authors reported no effect of N-POMC$_{1-76}$ on angiotensin II-, potassium-, or serotonin-stimulated aldosterone production by rat ZG cells. Pham-Huu-Trung *et al.* (1986) have also described a potentiation of ACTH action on aldosteronogenesis by γ$_3$-MSH (10^{-10} M) with normal human adrenal cells *in vitro*. Finally, Sharp and Sowers (1983) found that administration to spontaneously hypertensive and WKY rats of a specific anti-γ$_3$-MSH antiserum blocked the ACTH-mediated rise in plasma corticosterone, aldosterone, and 18-hydroxycorticosterone manifest in controls.

Studies from our laboratory and from that of Lowry have shown repeatedly that the presence of ACTH is required to elicit an effect of pro-γ-

MSHs on corticosterone or aldosterone secretion from normal rat or human adrenocortical cells. The observation by Jornot et al. (1985), therefore, that 16K fragment alone does not stimulate aldosterone secretion from rat ZG cells is not unexpected. Also, it is frequently overlooked that unlike the pro-γ-MSHs, γ-MSH itself is relatively inactive as an ACTH synergist (Pedersen et al., 1980).

While the studies cited above have established that pro-γ-MSHs can influence corticosteroidogenesis in human and rat adrenal tissue, there are indications that this may not be a universal phenomenon. Pham-Huu-Trung et al. (1982) found no significant steroidogenic activity for $γ_3$-MSH or 16K fragment on guinea pig adrenocortical cells, even in the presence of ACTH. We have confirmed this observation using guinea pig ZF cells and bovine $γ_3$-MSH (Ford-Holevinski and Brownie, unpublished observations). However, immunological data suggest that these results are explained by an apparent N-POMC sequence variation in this species (Pedersen and Brownie, unpublished observation).

2. Mechanism of Action of Pro-γ-MSHs

Apart from the observation that pro-γ-MSHs potentiate ACTH stimulation of aldosterone biosynthesis, there are no published data that directly address a mechanism of action. Problems confronting workers in this area include the limited amounts of tissue which can be harvested for experiments and the potential contamination from the adjoining ZF.

Nevertheless, we suggest that there may well be similarities between pro-γ-MSH action on the ZG and the way in which these polypeptides promote steroidogenesis in the inner zones of the adrenal cortex. If this is indeed the case, we can, for example, infer specific receptors for pro-γ-MSHs on the ZG plasma membrane comparable to those described for inner zone rat adrenal tissue (Pedersen and Brownie, 1983a). However, we would expect to find no effect of pro-γ-MSHs on ZG adenylate cyclase activity, for, in contrast to ACTH, treatment of rats with pro-γ-MSH does not activate either adenylate or guanylate cyclase (Pedersen and Brownie, 1983a) or increase intracellular cAMP levels (Farese et al., 1983) in the ZF/R. The identity of an intracellular second messenger for pro-γ-MSHs is therefore unresolved, but these data do clearly indicate that pro-γ-MSH, in contrast to α-MSH, is not simply a weak ACTH agonist.

It also remains to be demonstrated, but by analogy with tissue from the inner zones (Pedersen and Brownie, 1980; Pedersen et al., 1980) pro-γ-MSH may activate cytosolic cholesteryl ester hydrolase in the ZG. If this should prove to be correct, it would suggest that fluctuations in the rate of cholesteryl ester hydrolysis can modulate aldosterone secretion. Moreover, it could explain the observations of Cathiard et al. (1985). These investigators saw no effect of bovine γ_3-MSH or Lys-γ_3-MSH on gluco- and mineralocorticoid production by isolated bovine or ovine ZG and ZF/R cells, but the adrenals of these species have a markedly lower content of intracellular cholesteryl ester (Hechter, 1952; Glick and Ochs, 1955) than do the rat and human organs.

Although cholesterol side-chain cleavage in the ZG would seem to offer another potential site for control by pro-γ-MSHs, the enzyme complex in the ZF/R appears unresponsive to trypsinized 16K fragment (Pedersen and Brownie, 1980) or to γ_3-MSH (Pedersen et al., 1980). Moreover, since aldosterone biosynthesis in normal ZG cells is not independently stimulated by pro-γ-MSH, we can infer that the reaction is not regulated by this synergist. Thus, the steroidogenic effects of ACTH and of pro-γ-MSH in ZG cells should be qualitatively distinguishable from one another, as indeed they are in ZF/R cells.

Apart from the adrenotropic activity inherent in part of N-POMC, it will be of interest to see if polypeptides derived from the extreme N-terminal region of the molecule possess a mitogenic activity for ZG cells, as suggested for the adrenal as a whole by Lowry et al. (1983). Their provocative hypothesis implies that there are perhaps additional, more pervasive levels of N-POMC control than those considered here and that elevated concentrations of pro-γ-MSH and/or altered processing could promote chronic adrenocortical dysfunction.

3. Pro-γ-MSHs in Hyperaldosteronism

Lis et al. (1981) and Schiffrin et al. (1983) demonstrated that even in the absence of exogenous ACTH, human aldosteronoma cells in primary culture will respond to low concentrations of γ_3-MSH or N-POMC$_{1-76}$ (ED$_{50}$ $\cong 10^{-12}$–10^{-11} M) with an increased output of aldosterone. This contrasts sharply with the behavior of normal human adrenocortical cells described above. It suggests that in aldosteronoma cells the array of postreceptor mechanisms transducing the pro-γ-MSH signal has been altered in such a way that this hormone now elicits a more "ACTH-like" response from the tissue. One might hypothesize, for example, that as a consequence of transformation, the pro-γ-MSH membrane receptors and the adenylate cyclase system in these cells become coupled.

In collaboration with Melby and colleagues, we have confirmed these observations using dispersed human aldosteronoma cells and synthetic human γ_3-MSH (Aurecchia et al., 1982). In these incubations aldosterone secretion was more sensitive to γ_3-MSH than to ACTH, contrasting with the response of cortisol. Likewise, Pham-Huu-Trung et al. (1985) found that both N-POMC$_{1-76}$ and γ_3-MSH stimulated aldosteronoma cells at physiological concentrations (10^{-12}–10^{-10} M), potentiated ACTH action on aldosterone secretion, and increased the aldosterone/corticosterone ratio. Thus, these pro-γ-MSHs behave like angiotensin with respect to their preferential effect on mineralocorticoid secretion.

Despite these *in vitro* data, it is not yet certain that pro-γ-MSHs play a significant role in primary hyperaldosteronism. Güllner *et al.* (1983) failed to detect increased IR-γ-MSH in patients with either aldosterone-producing adenomas or IHA. The number of subjects, however, was small. In a study involving a larger number of patients, Griffing *et al.* (1985b) found that plasma IR-γ-MSH was above the normal range in those with aldosteronomas and was dramatically elevated in a subset of individuals with IHA. That finding, coupled with the observation that aldosteronoma cells are responsive to pro-γ-MSHs, suggests these polypeptides may function in the genesis of some forms of hyperaldosteronism.

Acknowledgments

The authors wish to acknowledge the expert secretarial assistance of Willy Brownie-Bakhuizen and Judith Colby. Original work from the authors' laboratory has been supported by National Institutes of Health research grants HL06975 and AM18141 to A. C. B. and a Research Career Development Award, HD00613, to R. C. P.

References

Aguilera, G., and Catt, K. J. (1979). *Endocrinology* **104**, 1046–1052.
Albano, J. D. M., Brown, B. L., Ekins, R. P., Tait, S. A. S., and Tait, J. F. (1974). *Biochem. J.* **142**, 391–400.
Al-Dujaili, E. A. S., Hope, J., Estivariz, F. E., Lowry, P. J., and Edwards, C. R. W. (1981). *Nature (London)* **291**, 156–159.
Anand-Srivastava, M. B., Genest, J., and Cantin, M. (1985). *FEBS Lett.* **181**, 199–202.
Andersen, J. M., and Dietschy, J. M. (1978). *J. Biol. Chem.* **253**, 9024–9032.
Andersen, J. M., and Dietschy, J. M. (1981). *J. Biol. Chem.* **256**, 7362–7370.
Andoka, G., Chauvin, M. A., Marie, J., Saez, J. M., and Morera, A. M. (1984). *Biochem. Biophys. Res. Commun.* **121**, 441–447.
Atarashi, K., Mulrow, P. J., Franco-Saenz, R., Snajdar, R., and Rapp, J. (1984). *Science* **224**, 992–994.
Atarashi, K., Mulrow, P. J., and Franco-Saenz, R. (1985). *J. Clin. Invest.* **76**, 1807–1811.
Aupetit, B., Accarie, C., Emeric, N., Vonarx, V., and Legrand, J.-C. (1983). *Biochim. Biophys. Acta* **752**, 73–78.
Aurecchia, S. A., Brownie, A. C., Pedersen, R. C., Raney, P., Allen, J., Griffing, G. T., and Melby, J. C. (1982). *Clin Res.* **30**, 489a.
Balla, T., Hunyady, L., and Spät, A. (1982). *Biochem. Pharmacol.* **31**, 1267–1271.
Balla, T., Enyedi, P., Hunyady, L., and Spät, A. (1984). *FEBS Lett.* **171**, 179–182.
Bell, J. B. G., Tait, J. F., Tait, S. A. S., Barnes, G. D., and Brown, B. L. (1981). *J. Endocrinol.* **91**, 145–154.

Braley, L. M., Menachery, A. I., and Williams, G. H. (1985). *Annu. Met. Am. Endocrinol. Soc., 67th* p. 207.
Brown, M. S., Kovanen, P. T., and Goldstein, J. L. (1979). *Recent Prog. Horm. Res.* **35**, 215–257.
Brown, R. D., Wisgerhof, M., Carpenter, P. C., Brown, G., Jiang, N.-S., Kao, P., and Hegstad, R. (1979). *J. Steroid Biochem.* **11**, 1043–1050.
Buckley, D. I., and Ramachandran, J. (1981). *Proc. Natl. Acad. Sci. U.S.A.* **78**, 7431–7435.
Buckley, D. I., Yamashiro, D., and Ramachandran, J. (1981). *Endocrinology* **109**, 5–9.
Campbell, W. B., and Gomez-Sanchez, C. E. (1985). *Endocrinology* **117**, 279–286.
Campbell, W. B., Currie, M. G., and Needleman, P. (1985). *Circ. Res.* **57**, 113–118.
Cathiard, A. M., Durand, Ph., Seidah, M. G., Chrétien, M., and Saez, J. M. (1985). *J. Steroid Biochem.* **23**, 185–190.
Chan, J. S. D., Lu, C.-L., Seidah, N. G., and Chrétien, M. (1982). *Endocrinology* **111**, 1388–1390.
Chan, J. S. D., Seidah, N. G., and Chrétien, M. (1983). *J. Clin. Endocrinol. Metab.* **56**, 791–796.
Chartier, L., and Schiffrin, E. L. (1986a). *Proc. Soc. Exper. Biol. Med.* **182**, 132–136.
Chartier, L., and Schiffrin, E. L. (1986b). *Fed. Proc.* **45**, 176.
Chartier, L., Schiffrin, E., Thibault, G., and Garcia, R. (1984). *Endocrinology* **115**, 2026–2028.
Davis, J. O. (1976). *Handb. Physiol. Endocrinol.* **6**, 77–106.
De Léan, A., Rácz, K., Gutkowska, J., Nguyen, T.-T., Cantin, M., and Genest, J. (1984). *Endocrinology* **115**, 1636–1638.
Douglas, J., Saltman, S., Williams, C., Bartley, P., Kondo, T., and Catt, K. (1978). *Endocr. Res. Commun.* **5**, 173–188.
Eipper, B. A., and Mains, R. E. (1978). *J. Biol. Chem.* **253**, 5732–5744.
Enyedi, P., Spät, A., and Antoni, F. A. (1981). *J. Endocrinol.* **91**, 427–437.
Enyedi, P., Büki, B., Musci, I., and Spät, A. (1985). *Mol. Cell. Endocrinol.* **41**, 105–112.
Fakunding, J. L., and Catt, K. J. (1980). *Endocrinology* **107**, 1345–1353.
Fakunding, J. L., Chow, R., and Catt, K. J. (1979). *Endocrinology* **105**, 327–333.
Farese, R. V., Ling, N. C., Sabir, M. A., Larson, R. E., and Trudeau, W. L. (1983). *Endocrinology* **112**, 129–132.
Fujita, K., Aguilera, G., and Catt, K. J. (1979). *J. Biol. Chem.* **254**, 8567–8574.
Gallo-Payet, N., and Escher, E. (1985). *Endocrinology* **117**, 38–46.
Glick, D., and Ochs, M. J. (1955). *Endocrinology* **56**, 285–298.
Glickman, J. A., Carson, G. D., and Challis, J. R. G. (1979). *Endocrinology* **104**, 34–39.
Goodfriend, T. L., Elliott, M. E., and Atlas, S. A. (1984). *Life Sci.* **35**, 1675–1682.
Griffing, G. T., McIntosh, T., Berelowitz, B., Hudson, M., Salzman, R., Manson, J.-A. E., and Melby, J. C. (1985a). *J. Clin. Endocrinol. Metab.* **60**, 315–319.
Griffing, G. T., Berelowitz, B., Hudson, M., Salzman, R., Manson, J.-A. E., Aurrechia, S., Melby, J. C., Pedersen, R. C., and Brownie, A. C. (1985b). *J. Clin. Invest.* **76**, 163–169.
Gross, M. D., Grekin, R. J., Brown, L. E., Marsh, D. D., and Beierwaltes, W. H. (1981). *J. Clin. Endocrinol. Metab.* **52**, 612–615.
Güllner, H.-G., and Gill, J. R. (1983). *J. Clin Invest.* **71**, 124–128.
Güllner, H.-G., Nicholson, W. E., Gill, J. R., and Orth, D. N. (1983). *J. Clin. Endocrinol. Metab.* **56**, 853–855.
Gwynne, J. T., and Strauss, J. F., III (1982). *Endocr. Rev.* **3**, 299–329.
Hale, A. C., Ratter, S. J., Tomlin, S. J., Lytras, N., Besser, G. M., and Rees, L. H. (1984). *Clin. Endocrinol.* **21**, 139–148.

Haning, R., Tait, S. A. S., and Tait, J. F. (1970). *Endocrinology* **87**, 1147–1167.
Hechter, O. M. (1952). *In* "Adrenal Cortex: Transactions of the Third Conference" (E. P. Ralli, ed.), pp. 120–121. Progress Associates, Caldwell, New Jersey.
Hinson, J. P., Vinson, G. P., Whitehouse, B. J., and Price, G. (1985). *J. Endocrinol.* **104**, 387–395.
Hyatt, P. J., Bell, J. B. G., Bhatt, K., Chu, F. W., Tait, J. F., Tait, S. A. S., and Whitley, G. St.J. (1985). *J. Endocrinol.* **107** (Suppl.), Abstr. 32.
Jornot, L. H., Capponi, A. M., and Vallotton, M. B. (1985). *J. Steroid Biochem.* **22**, 221–225.
Kaplan, N. M. (1965). *J. Clin. Invest.* **44**, 2029–2039.
Kaplan, N. M., and Bartter, F. C. (1962). *J. Clin. Invest.* **41**, 715–724.
Kater, C. E., Biglieri, E. G., Rost, C. R., Schambelan, M., Hirai, J., Chang, B. C. F., and Brust, N. (1985). *J. Clin. Endocrinol. Metab.* **60**, 225–228.
Kem, D. C., Feldman, M., Starkweather, G., and Li, C. H. (1985). *J. Clin. Endocrinol. Metab.* **60**, 440–443.
Kojima, I., Inano, H., and Tamaoki, B.-I. (1982). *Biochem. Biophys. Res. Commun.* **106**, 617–624.
Kojima, I., Ogata, E., Inano, H., and Tamaoki, B.-I. (1984a). *Acta Endocrinol. (Copenhagen)* **107**, 395–400.
Kojima, I., Kojima, K., Kreutter, D., and Rasmussen, H. (1984b). *J. Biol. Chem.* **259**, 14448–14457.
Kojima, I., Kojima, K., and Rasmussen, H. (1985a). *J. Biol. Chem.* **260**, 4248–4256.
Kojima, I., Kojima, K., and Rasmussen, H. (1985b). *J. Biol. Chem.* **260**, 9171–9176.
Kramer, R. E., Gallant, S., and Brownie, A. C. (1979). *J. Biol. Chem.* **254**, 3953–3958.
Kramer, R. E., Gallant, S., and Brownie, A. C. (1980). *J. Biol. Chem.* **255**, 3442–3447.
Krieger, D. T., Liotta, A., and Li, C. H. (1977). *Life Sci.* **21**, 1771–1777.
Kudo, T., and Baird, A. (1985). *Nature (London)* **312**, 756–757.
Lee, T. C., and de Wied, D. (1968). *Life Sci.* **7**, 35–45.
Lee, T. C., van der Wal, B., and de Wied, D. (1968). *J. Endocrinol.* **42**, 465–475.
Li, C. H., Ng, T. B., and Cheng, C. H. K. (1982). *Int. J. Pept. Protein Res.* **19**, 361–365.
Lieberman, A. H., and Luetscher, J. A. (1960). *J. Clin. Endocrinol.* **20**, 1004–1016.
Ling, N., Ying, S., Minick, S., and Guillemin, R. (1979). *Life Sci.* **25**, 1773–1780.
Lis, M., Hamet, P., Gutkowska, J., Maurice, G., Seidah, N. G., Larivière, B., Chrétien, M., and Genest, J. (1981). *J. Clin. Endocrinol. Metab.* **52**, 1053–1056.
Llanos, A. J., Ramachandran, J., Creasy, R. K., Rudolph, A. M., and Serón-Ferré, M. (1979). *Endocrinology* **105**, 613–617.
Lowry, P. J., and McMartin, C. (1972). *J. Endocrinol.* **55**, xxxiii.
Lowry, P. J., Silas, L., McLean, C., Linton, E. A., and Estivariz, F. E. (1983). *Nature (London)* **306**, 70–73.
Lymangrover, J. R., Keku, E., and Eldridge, J. C. (1983). *Life Sci.* **33**, 1605–1612.
McCaa, R. E., Young, D. B., Guyton, A. C., and McCaa, C. S. (1974). *Circ. Res.* **34–35** (Suppl. I), 15–25.
McCaa, R. E., Montalvo, J. M., and McCaa, C. S. (1978). *J. Clin. Endocrinol. Metab.* **46**, 247–254.
McLoughlin, L., Lowry, P. M., Ratter, S., Besser, G. M., and Rees, L. H. (1980). *Clin. Endocrinol.* **12**, 287–292.
Marusic, E. T., White, A., and Aedo, A. R. (1973). *Arch. Biochem. Biophys.* **157**, 320–321.
Matsuoka, H., Mulrow, P. J., and Li, C. H. (1980a). *Science* **209**, 307–308.
Matsuoka, H., Mulrow, P. J., Franco-Saenz, R., and Li, C. H. (1980b). *Clin. Sci.* **59**, 91s–94s.

Matsuoka, H., Mulrow, P. J., Franco-Saenz, R., and Li, C. H. (1981a). *Science* **291**, 155–156.
Matsuoka, H., Mulrow, P. J., Franco-Saenz, R., and Li, C. H. (1981b). *J. Clin Invest.* **68**, 752–759.
Miller, R. T., Douglas, J. G., and Dunn, M. J. (1980). *Prostaglandins* **20**, 449–462.
Motomatsu, T., Takahashi, H., Ibayashi, H., and Nobunaga, M. (1984). *J. Clin. Endocrinol. Metab.* **59**, 495–498.
Müller, J. (1966). *Acta Endocrinol. (Copenhagen)* **52**, 515–526.
Nagy, K., Koroknai, L., and Spát, A. (1984). *J. Steroid Biochem.* **20**, 789–791.
Nakanishi, S., Inoue, A., Kita, T., Nakamura, M., Chang, A. C. Y., Cohen, S. N., and Numa, S. (1979). *Nature (London)* **278**, 423–427.
Nakao, K., Nakai, Y., Oki, S., Horii, K., and Imura, H. (1978). *J. Clin. Invest.* **62**, 1395–1398.
Page, R. B., Boyd, J. E., and Mulrow, P. J. (1974). *Endocr. Res. Commun.* **1**, 53–62.
Palmore, W. P., and Mulrow, P. J. (1967). *Science* **158**, 1482–1484.
Palmore, W. P., Anderson, R., and Mulrow, P. J. (1970). *Endocrinology* **86**, 728–734.
Pedersen, R. C., and Brownie, A. C. (1980). *Proc. Natl. Acad. Sci. U.S.A.* **77**, 2239–2243.
Pedersen, R. C., and Brownie, A. C. (1983a). *Endocrinology* **112**, 1279–1287.
Pedersen, R. C., and Brownie, A. C. (1983b). *Proc. Natl. Acad. Sci. U.S.A.* **80**, 1882–1886.
Pedersen, R. C. and Brownie, A. C. (1986). *In* "Biochemical Actions of Hormones" (G. Litwack, ed.), Vol. XIII, pp. 129–166. Academic Press, Orlando.
Pedersen, R. C., Brownie, A. C., and Ling, N. (1980). *Science* **208**, 1044–1045.
Pedersen, R. C. Ling, N., and Brownie, A. C. (1982). *Endocrinology* **110**, 825–834.
Peytremann, A., Brown, R. D., Nicholson, W. E., Island, D. P., Liddle, G. W., and Hardman, J. G. (1974). *Steroids* **24**, 451–462.
Pham-Huu-Trung, M.-T., De Smitter, N., Bogio, A., Bertagna, X., and Girard, F. (1982). *Endocrinology* **110**, 1819–1821.
Pham-Huu-Trung, M.-T., Bogyo, A., de Smitter, N., and Girard, F. (1985). *J. Clin. Endocrinol. Metab.* **61**, 467–471.
Pham-Huu-Trung, M.-T., Bogyo, A., Leneuve, P., and Girard, F. (1986). *J. Steroid Biochem.* **24**, 345–348.
Rabinowe, S. L., Taylor, T., Dluhy, R. G., and Williams, G. H. (1985). *J. Clin. Endocrinol. Metab.* **60**, 485–489.
Rácz, K., Gláz, E., Kiss, R., Lada, G., Varga, I., Vida, S., di Gleria, K., Medzihradszky, K., Lichtwald, K., and Vecsei, P. (1980). *Biochem. Biophys. Res. Commun.* **97**, 1346–1353.
Rácz, K., Varga, I., Gláz, E., Kiss, R., Vida, S., Lada, G., di Gleria, K., Medzihradszky, K., Lichtwald, K., and Vecsei, P. (1982). *J. Clin. Endocrinol. Metab.* **54**, 656–660.
Rácz, K., Kuchel, O., Cantin, M., and De Léan, A. (1985). *FEBS Lett.* **192**, 19–22.
Rapp, J. P., and Dahl, L. K. (1976). *Biochemistry* **15**, 1235–1242.
Saruta, T., and Kaplan, N. M. (1972). *J. Clin. Invest.* **51**, 2246–2251.
Saruta, T., Cook, R., and Kaplan, N. M. (1972). *J. Clin. Invest.* **51**, 2239–2245.
Schiffrin, E. L., Chrétien, M., Seidah, N. G., Lis, M., Gutkowska, J., Cantin, M., and Genest, J. (1983). *Horm. Metab. Res.* **15**, 181–184.
Seelig, S., and Sayers, G. (1973). *Arch. Biochem. Biophys.* **154**, 230–239.
Sen, S., Bravo, E. L., and Bumpus, F. M. (1977). *Circ. Res.* **40**, (Suppl. I), I5–I10.
Sen, S., Shainoff, J. R., Bravo, E. L., and Bumpus, F. M. (1981a). *Hypertension* **3**, 4–10.
Sen, S., Valenzuela, R., Smeby, R., Bravo, E. L., and Bumpus, F. M. (1981b). *Hypertension* **3** (Suppl. 1), I81–I86.

Sharp, B., and Sowers, J. R. (1983). *Biochem. Biophys. Res. Commun.* **110**, 357–363.
Shenker, Y., Villareal, J. Z., Sider, R. S., and Grekin, R. J. (1985). *Endocrinology* **116**, 138–141.
Silman, R. E., Chard, T., Lowry, P. J., Smith, I., and Young, I. M. (1976). *Nature (London)* **260**, 716–718.
Solyom, J., Ludwig, E., and Vajda, A. (1971). *Acta Physiol. Acad. Sci. Hung.* **39**, 343–349.
Swaab, D. F., and Visser, M. (1977). *Front. Horm. Res.* **4**, 170–178.
Szabo, D., Sazlay, K. Sz., and Toth, I. E. (1984). *Mol. Cell. Endocrinol.* **34**, 59–66.
Szalay, K. Sz., and Stark, E. (1981). *Life Sci.* **29**, 1355–1361.
Szalay, K. Sz., and Stark, E. (1982). *Life Sci.* **30**, 2101–2108.
Tanaka, K., Nicholson, W. E., and Orth, D. N. (1977). *Front. Horm. Res.* **4**, 208–214.
Tanaka, K., Nicholson, W. E., and Orth, D. N. (1978). *J. Clin. Invest.* **62**, 94–104.
Thody, A. J., Celis, M. E., and Fisher, C. (1980). *Peptides* **1**, 125–129.
Thody, A. J., Fisher, C., Kendal-Taylor, P., Jones, M. T., Price, J., and Abraham, R. R. (1985). *Acta Endocrinol.* 313–318.
Tremblay, J., Gerzer, R., Pang, S. C., Cantin, M., Genest, J., and Hamet, P. (1986). *FEBS Lett.* **194**, 210–214.
Ulick, S. (1972). *Proc. Int. Congr. Endocrinol., 4th, Excerpta Medica Int. Congr. Ser.* **273**, 761.
Ulick, S. (1976). *J. Clin. Endocrinol. Metab.* **43**, 92–96.
van der Wal, B., and de Wied, D. (1968). *Acta Endocrinol. (Copenhagen)* **59**, 186–192.
Vinson, G. P., Whitehouse, B. J., Dell, A., Etienne, A. T., and Morris, H. R. (1980). *Nature (London)* **284**, 464–467.
Vinson, G. P., Whitehouse, B. J., Dell, A., Etienne, A. T., and Morris, H. R. (1981a). *Biochem. Biophys. Res. Commun.* **99**, 65–72.
Vinson, G. P., Whitehouse, B. J., and Thody, A. J. (1981b). *Peptides* **2**, 141–144.
Vinson, G. P., Whitehouse, B. J., Dell, A., Bateman, A., and McAuley, M. E. (1983). *J. Steroid Biochem.* **19**, 537–544.
Vinson, G. P., Whitehouse, B. J., Bateman, A., Hruby, V. J., Sawyer, T. K., and Darman, P. S. (1984). *Life Sci.* **35**, 603–610.
Waldman, S. A., Rapoport, R. M., Fiscus, R. R., and Murad, F. (1985). *Biochim. Biophys. Acta* **845**, 298–303.
Washburn, D. D., Kem, D. C., Orth, D. N., Nicholson, W. E., Chretién, M., and Mount, C. D. (1982). *J. Clin. Endocrinol. Metab.* **54**, 613–618.
Williams, G. H., Rose, L. I., Dluhy, R. G., Dingman, F., and Lauler, D. P. (1971). *J. Clin. Endocrinol. Metab.* **32**, 27–35.
Wilson, J. F., and Harry, F. M. (1980). *J. Endocrinol.* **86**, 61–67.
Wilson, J. F., and Morgan, M. A. (1979). *J. Endocrinol.* **82**, 361–366.
Wilson, J. X., Aguilera, G., and Catt, K. J. (1984). *Endocrinology* **115**, 1357–1363.
Yamakado, M., Franco-Saenz, R., and Mulrow, P. J. (1985). *Proc. Soc. Exp. Biol. Med.* **179**, 318–323.
Yanagibashi, K., Haniu, M., Shively, J. E., Shen, W. H., and Hall, P. (1986). *J. Biol. Chem.* **261**, 3556–3562.

6

Behavioral Actions of ACTH and Related Peptides

CURT A. SANDMAN* AND ABBA J. KASTIN†

*Department of Psychiatry and Human Behavior
Department of Psychobiology
University of California, Irvine
State Developmental Research Institutes
Fairview Developmental Center
Costa Mesa, California 92626
and
† Veterans Administration Medical Center and
Tulane University School of Medicine
New Orleans, Louisiana 70146

I. Introduction

Numerous recent reviews of the actions of neuropeptides exist in the literature (e.g., see Bertolini and Gessa, 1981; DeWied, 1983; DeWied and Jolles, 1982; DeWied and van Ree, 1982; Kastin et al., 1976, 1980, 1981, 1983, 1984; Sandman and Kastin, 1981a,b; Sandman et al., 1977b, 1981; Sandman and O'Halloran, 1986; Tinklenberg and Thorton, 1983). Of course this reflects the ever-increasing number of empirical studies and the escalating interest in this unique class of neurotransmitter. In this review we will attempt to present a comprehensive analysis of the behavioral actions of ACTH and its family of fragments and related molecules, specifically the melanocyte-stimulating hormones (MSHs). Although the independent actions of ACTH/MSH will be emphasized (Kastin et al., 1981), it is acknowledged that these neuropeptides are derived from a larger structure, proopiomelanocortin (POMC), that contains several behaviorally interesting fragments. The fragments within POMC may define a dynamic, self-regulating network. It has been proposed that either reciprocal (Sandman and Kastin, 1981a,b) or redundant (DeWied et al., 1978) properties characterize the relationship among fragments of POMC. Far more relevant to this chapter is the fact that the α-MSH sequence is duplicated in the ACTH molecule. Thus, although we shall focus on the behavioral action of ACTH/MSH, it is with full recognition of the incompletely defined dynamic relationships that determine their functional significance.

POMC is a common precursor for several behaviorally active neuropeptides (peptides found in the nervous system), including α-, β-, and γ-MSH, ACTH, and β-LPH and the family of endorphins. POMC is synthesized in the anterior and intermediate lobes of the pituitary as well as in extrapituitary sites such as the amygdala and hypothalamus (Civelli et al., 1982), the human adrenals (Evans et al., 1983), the gastrointestinal tract, lungs, and pancreas (Krieger et al., 1980). These findings may be important in understanding the behavioral actions of POMC fragments since subcutaneous injections of these peptides have been found to influence physiological mechanisms and behavioral expression in humans.

Of more direct interest for behavior is the distribution of POMC-derived neuropeptides in "cognitive-relevant" brain regions such as hippocampus and cortex (Rudman et al., 1974; Eskay et al., 1979; Borvendez et al., 1978; Gramsch et al., 1980; Krieger et al., 1977; Moldow and Yalow, 1978). The origin of brain-localized POMC fragments is uncertain. Krieger et al. (1980) reviewed strong evidence indicating that the arcuate nucleus is the site of origin of cell bodies for a large proportion of at least ACTH and β-endorphin fibers. Similarly, transectional isolation of the hypothalamic arcuate nucleus depletes brain tissue of α-MSH (Eskay et al., 1979). A second neuronal system may arise in the dorsal lateral hypothalamus (Piekut and Knigge, 1984). The contribution of the pituitary lobes to brain POMC fragments presently is uncertain, and until recently the contribution of the pituitary was restricted to release into the general circulation for peripheral target sites. However, there have been indications of retrograde transport of POMC peptides back into the brain following intrapituitary administration (Dorsa et al., 1982; Mezey and Palkovits, 1982). In light of the very high concentrations of pituitary peptides in hypophyseal effluent (100–1000 times greater than in peripheral blood), retrograde transport may provide an important avenue of delivery for POMC products to brain.

II. Stress

Changes in the environment are reflected by changes in the nervous system. Our ability to respond to, and remember, changing environmental events implicate a tight bond between the environment and the organism. Some changes are instantaneous while others are gradual. Changes in the nervous system occur against a background of activity, or a host resistance (Veith and Sandman, 1985), which sets the limits of response. The many functions described for ACTH may be predicted by the dynamic range of changes in response to, or preparation for, environmental challenge. Among the oldest conception of the actions of ACTH is its role in the stress response (Selye, 1936, 1956).

Selye's general-adaptation syndrome (GAS) was conceived as a homeostatic, emergency response to changes in the environment. Paramount in his conception was the release of ACTH from the pituitary during stress that triggered the release of adrenocortical steroids from the adrenal gland to prepare the organism to cope with infection and trauma. Obviously, this putative invariant pattern was adaptive. However, if overused, as in the case of chronic stress, it became exhausted, rendering the organism vulnerable to disease. Psychosomatic medicine arose from Selye's forceful analysis and currently is enjoying a renewal as behavioral medicine and psychoimmunology.

Even though some of the early tenets of Selye's theory appear incorrect, the modifications added support to his general thesis. For example, current results indicate a tight coupling between specific environmental events and discrete patterns of endocrine response (Mason, 1975). Thus, the initial conception of a general effect of stress on ACTH release has been replaced by current theories emphasizing highly specific response patterns to the environment. Perhaps nowhere in the neuropeptide field has the specificity idea been explored more actively than with ACTH and behavior.

A. GROOMING

One manifestation of increased anxiety or stress in animals is grooming behavior (Williams and Scampoli, 1984). As extensively reviewed by Gispen and Isaacson (1981), ACTH (1–24) and (1–16) extend both stress- and novelty-induced grooming in the rat. Intraventricular injections of ACTH (1–24), (1–16), (1–23), (5–18), and (5–16), but not (1–10), (4–10), (11–24), or (7–16), initiate grooming in the nonstressed rat (Gispen et al., 1976; Gispen, 1982). Grooming in the presence of ACTH only can be suppressed by more basic drives such as hunger or thirst (Jolles et al., 1979). Peptides that release ACTH, such as corticotropin releasing factor, sauvagine, and urotensin I, also induce grooming in the rat (Britton et al., 1984). Interestingly, handling and the opiate antagonist naloxone also induce grooming (Williams and Scampoli, 1984). Since β-endorphin induces grooming, the similar effect of naloxone is puzzling. However, Gispen and Isaacson (1981) suggest that β-endorphin-related grooming is qualitatively different than that stimulated by ACTH in the sense that the opiates promote general and diffuse activation.

B. Stretching–Yawning Syndrome (SYS)

Among the earliest observations of the behavioral influence of ACTH was the stretching–yawning syndrome (SYS) after intracisternal injection of ACTH into dogs (Ferrari et al., 1955, 1963). The SYS is thought to be a vestige of adaptive mechanisms to antagonize sleep under dangerous circumstances (Bertolini and Gessa, 1981). The MSH configuration is the optimal peptide sequence for eliciting the syndrome, conceivably indicating its resemblance to camouflage in amphibians. The SYS has been expanded to include sexual arousal, since penile erection (Genedani et al., 1984) and lordosis (Thody and Wilson, 1983) also have been observed after intraventricular injections. Typically, the SYS is observed only after central administration (Bertolini et al., 1975) with areas of the hypothalamus and lining of the third ventricle involved in the ACTH-induced effect (Gessa et al., 1967). Thody and Wilson (1983) reported that peripheral injections inhibited lordosis. They speculated two effects, a central effect controlled by the arcuate nucleus, preoptic nucleus, and midbrain that reflected arousal, and a peripheral effect controlled by the adrenal that was inhibitory.

Among the most provocative studies was the recent report of Genedani et al. (1984) These authors conjectured that the ACTH-induced SYS and sexual arousal were due to the role of polyamines and protein synthesis in the brain. Since ACTH stimulates brain ornithine decarboxylase (ODC) activity and increases putrescine (a polyamine), blockade of these actions may inhibit the ACTH-induced effect. Indeed, pretreatment of rats with α-difluoromethyl ornithine (DEMO), which blocks ODC, strongly inhibited ACTH-induced SYS and, especially, penile erection. Polyamines when injected directly do not produce SYS; they may act to stabilize membranes for secure ACTH–receptor coupling or inhibit ACTH-induced synthesis of new proteins. Whether these mechanisms also operate for the cognitive influences of ACTH remains an interesting question for future research.

C. Social Behavior

Rats housed in isolation show increased social (proximity between animals) interaction when tested in pairs or small groups. The effects of ACTH-like peptides have been reported to increase (Beckwith et al., 1977a), decrease (Niesink and van Ree, 1984), have no effect (Crawley et al., 1981; Crabbe et al., 1982), or to normalize (Niesink and van Ree, 1983) social behavior. Slightly different test conditions were apparent in these studies, indicating that this behavioral response is fragile. For instance, the study by Beckwith et al. involved treating the animals with peptides as infants and testing them at 45 and 120 days of age. The results were complex, indicating effects of both time of testing and sex of animal (the effect persisted longer in males but was stronger in females). However, the normalizing influence of ACTH on social behavior was the most interesting.

In a well-conceived study (Niesink and van Ree, 1983), rats were tested after short-term isolation in either a familiar or unfamiliar environment with intense or dim light. Pretreatment with the ACTH (4–9) analog decreased social interactions due to short-term isolation but increased the social contact provoked by novel surroundings. That is, the "normal" social behavior in response to the stress of isolation and novelty was blocked by the ACTH analog. In a companion study, these authors (Niesink and van Ree, 1984) reported that ACTH (1–24), (4–10), and (1–13) were not effective, but only the (4–9) analog and α-MSH had what they termed "normalizing effects." They speculated that these effects were mediated by interaction with the endogenous opiate system.

III. Opiate-Like Behavioral Effects

As with social behavior, there are conflicting reports about the opiate-like effects of ACTH and related peptides. A variety of opiate-like behaviors has been investigated using different procedures and doses of peptides, accounting for the apparent inconsistency in the literature. As reviewed by Bertolini and Gessa (1981) and Mousa and Couri (1983), only intraventricular injection of ACTH results in hyperalgesia (increased experience of pain). However, among the earliest reports (Winter and Flataker, 1951) were those of peripheral injections of ACTH inducing hyperalgesia and reversing the effects of morphine. Later, Sandman and Kastin (1981b) and Bertolini et al. (1980) reported hyperalgesia after ICV injections of MSH and ACTH, respectively. The unusual finding in the study of Sandman and Kastin was the small dose (1 μg) injected to produce sustained hyperalgesia as compared with the doses of Bertolini et al. (1979), which were 20–50 μg. Large doses of the ACTH (4–9) analog (30 μg) injected directly into the midbrain produced significant analgesia (Walker et al., 1981). This finding is consistent with the results of Jacquet and Wolf (1981) involving injections of ACTH into the periaqueductal gray that induced the opiate abstinence syndrome. These effects were similar to the excitatory, but not the inhibitory, action of morphine. Jacquet and Wolf (1981) have proposed that there is a second class of opiate receptors for which ACTH is an endogenous ligand.

At least two studies have compared γ-MSH (Tyr-Val-Met-Gly-His-Phe-Arg-Trp-Asp-Arg-Phe-Gly) with ACTH-like peptides (O'Donohue et al., 1981a; van Ree et al., 1981). Both studies found that the effects of injections of γ-MSH were opposite to those of the ACTH/MSH peptide. Van Ree et al. (1981) indicated that γ-MSH resembled opiate antagonists and labeled it a partial ACTH agonist/antagonist. Specifically, these researchers reported that γ-MSH displaced naloxone binding, had a modest effect on ileal contractions, and slightly attenuated ICV β-endorphin analgesia. The naloxone-like effects were more evident at higher doses.

The search for endogenous opiate antagonists in addition to Try-MIF-1 (Tyr-Pro-Leu-Gly-NH_2) continues. The findings for ACTH/MSH and γ-MSH indicate they may be candidates. Since the effects of simultaneous administration of γ-MSH and α-MSH have not been additive (O'Donohue et al., 1981a), the possibility of a polyreceptor ligand from these candidates within the POMC precursor molecule appears unlikely.

IV. Learning, Attention, and Memory

A. Avoidance Conditioning

Many approaches have been used to examine the effects of ACTH on learning and memory, but the active avoidance paradigm, rooted in the stress–fear tradition, remains the most frequently used measure. Typically, in an active avoidance paradigm, an animal is presented with a neutral stimulus [conditioned stimulus (CS)] that precedes shock [unconditioned stimulus (UCS)] by a few seconds. The animal must learn to respond to the CS (e.g., run to another chamber, jump on a pole) in order to avoid the UCS. During extinction, the UCS is turned off and the persistence of the animal's response is interpreted as an index of memory. It is generally assumed that the neutral stimulus acquires its motivational significance by its association with shock and makes the animal fearful. Thus, the animal learns to avoid and subsequently remembers the CS because it elicited fear. In a passive avoidance paradigm, the animal must learn to associate and remember that punishment is paired with a desirable response (e.g., moving into a darkened chamber). The animal is tested for its ability to inhibit a response. Since the ACTH response in response to shock was considered a manifestation of the fear, the extrapolation that injection of ACTH improved avoidance conditioning because it heightened the animal's own fearfulness, dominated scientific thinking for at least 15 years.

The early research of R. Miller and his associates indicated that treatment of intact and adrenalectomized rats with the entire ACTH chain significantly prolonged extinction of a learned avoidance response without influencing acquisition (Murphy and Miller, 1955; Miller and Ogawa, 1962). The initial departure from the classical fear-or-stress hypothesis was proposed by DeWied and Bohus (1966). They interpreted the prolonged extinction observed after treatment with MSH/ACTH as indicating that these peptides had neurotropic effects and enhanced memory processes directly. Although it was stated in their early reports that trial-to-trial memory was primarily influenced, subsequently they argued that enhancement of memory was a function of the increased general motivational state of the organism. Nevertheless, several experiments have implicated MSH/ACTH fragments directly in retrieval processes. Rigter and co-workers (Rigter, 1978; Rigter and van Riezen, 1975; Rigter et al., 1976) trained animals to passive avoid shock and then induced amnesia by applying electroconvulsive shock or partial asphyxiation by CO_2. These treatments erase the memory of the learning experience. However, treatment with MSH/ACTH (4–10) before the test of retention restored the memory of the experience. These results provided strong support for the proposal that short-term memory processes, especially those involved in retrieval of information from long-term memory storage, were facilitated by treatment with MSH/ACTH fragments. Consistent with this reasoning are the findings that ACTH facilitated escape from aversive stimuli (Mirsky et al., 1953), because in these aversive learning paradigms, ACTH "amplifies" the significance of stimuli (Stratton and Kastin, 1974).

However, several factors cloud an unequivocal acceptance of the memory hypothesis. First, there are differences in the avoidance conditioning performance depending on whether the injections were central or systemic. de Almeida et al. (1983) reported that systemic injections of either ACTH or epinephrine delayed extinction of a passive avoidance response. However, intraventricular injections had no effect. They argued, as has McGaugh (1983), that these compounds may influence memory through peripherally mediated routes. There are, however, a number of studies that has implicated central processes (Vecesi et al., 1981) in the effects of ACTH [see DeWied and Jolles (1982) for a thorough review]. The second issue relates to whether or not ACTH and its fragments have identical actions. The findings that ACTH can restore avoidance conditioning after hypophysectomy, but that ACTH (4–10) is less effective (Weiss et al., 1970; DeWied, 1964, 1967), argue against the belief that completely redundant information is contained in these peptides (Sandman and Kastin, 1981a; Sandman *et al.,* 1980a; Drago *et al.,* 1984). Thus, even though some fragments may influence memory, some may not. Third, some studies found that ACTH given just before retention augments memory (van Wimersma Griedanus *et al.,* 1978; Rigter and van Riezen, 1975; Rigter *et al.,* 1976), whereas others have found maximal enhancement when injections were coupled with the acquisition phase (Gold and van Buskirk, 1976a,b). Last, the effect of dose on avoidance conditioning echoes a consistent theme; low doses enhance memory and high doses disrupt memory (Gold and van Buskirk, 1976a,b). This issue is even more salient in studies of human subjects reviewed below.

B. Visual Discrimination and Reversal Learning

Among the most robust cognitive effects of ACTH/MSH in rats is improved reversal learning. With this procedure, animals are trained to either avoid shock or attain a reward by learning to discriminate binary cues (such as light–dark, black–white). In the initial phase of testing, the animal is rewarded for choosing the white door or lighted alley of a test apparatus. After acquiring the response, the animal must learn the opposite response (the black door or the darkened alley). This phase is called reversal learning.

According to MacKintosh (1965, 1969), the reversal problem is a measure of the selective attentional capacity of the animal. During original learning, the animal learns about the dimension of brightness as well as the specific response (i.e., to choose the white door). When the problem is reversed, the attentive animal tests values on the selected dimension of brightness (black–white) rather than irrelevant dimensions (e.g., spatial localization, left–right) because it has learned to resist distracting information. The effects of overtraining during initial learning were marshaled as evidence for the attentional explanation. For example, rats given as many as 500 trials beyond the learning criterion in the initial phase solve the reversal problem faster (i.e., they give up the old habit faster) than animals not given overtraining. This counterintuitive finding is difficult to reconcile with common sense and the classical view that, as habit strength increases and as learning proceeds, overlearned information would be resistant to new learning. The suggestion that another process, such as attention, is strengthened during overtraining was offered as an explanation of the finding, under these circumstances, that reversal learning is enhanced. In a series of studies designed to examine the various psychological constructs proposed to account for the influence of neuropeptides on behavior (Sandman *et al.,* 1972, 1973, 1974, 1980a), rats were trained with this two choice, simultaneous visual discrimination problem to avoid shock by running to a white door. After the animals acquired the response and avoided the shock, the task was reversed so that the simultaneously available black door was the correct response. As reviewed above, the first stage of the experiment measured the animal's ability to learn a new response. The reversal stage measured the animal's selective attention.

Treatment of rats with MSH had no appreciable effect on original learning. However, rats treated with MSH during the initial problem required approximately 50% fewer trials to solve the reversal learning problem. This effect was strongest when animals were tested in conditions that interfered with optimal performance (Sandman et al., 1972, 1973). Thus, injections of the neuropeptide MSH/ACTH created a selective attentional set that paralleled many trials of overlearning. These findings have been extended by Landfield et al. (1981) to aged animals. These authors reported that ACTH fragments may preserve, selectively, the ability of elderly rats to perform the visual reversal learning problem. In addition, the reports of O'Donohue et al. (1981b, 1982) indicated that the deaminated form of α-MSH may be especially potent in enhancing the ability of rats to perform a reversal task. We have concluded from these results, as well as from data gathered in other paradigms, that the MSH/ACTH peptide enhanced attentional processes.

In a recent, refined analysis, we (Sandman et al., 1980a) compared the influence of the family of MSH/ACTH peptides, MSH/ACTH (4–10), α-MSH (1–13), β-MSH (1–18), β-MSH (1–22), and ACTH (1–24), on discrimination and attention. This study was designed to evaluate the influence of the redundant chemical information stored in these related peptide chains. Although the prevailing view of these structure–activity relationships was that behavioral information in these molecules was redundant, Greven and DeWied (1977) have indicated that the proposed redundancy may have been restricted to extinction of the pole-jumping avoidance response.

The results of our study indicated that the rate of learning the original problem increased with administration of compounds of increasing molecular weight. The initial stage of learning was enhanced significantly with administration of MSH/ACTH (4–10). Except for ACTH (1–24) (the only peptide in this group that stimulates release of adrenal steroids), all of the other peptides also improved learning, though not achieving acceptable levels of statistical significance.

The structure–activity relationships were much different for reversal learning. Maximal enhancement of reversal learning (an index of attention) was achieved with administration of α- and β-MSH. Thus, when plotted according to molecular weight, a significant quadratic relationship with reversal learning (attention) was apparent.

The results of the early phases of the learning process (original learning) extended the conclusions of DeWied and Bohus (1966) and Greven and DeWied (1977). It appeared that the redundancy proposed for MSH/ACTH fragments was not restricted to a pole-jumping avoidance task but included, as well, the behaviors of learning a visual discrimination problem. It is tempting to speculate that behaviors that may be enhanced by nonspecific arousal share a monotonic relationship between performance and molecular weight. The relationships observed in this study supported such a speculation and suggested that trial-to-trial memory may be influenced by the (4–10) fragment.

The relationship of α-MSH, especially in its deaminated form, is illustrated further by a series of elegant studies by O'Donohue and colleagues (1981a, 1982). Both α-MSH and N-deacetylated MSH are found in the rat and human brain. However, these similar peptides have strikingly different influences on learning. α-MSH produced a 100% saving, whereas the deacetylated form had no influence on reversal learning (O'Donohue et al., 1981a). Recently (Kobobun et al., 1983) this group reported that an MSH analog (N-leucine 4-D-Phe-7), with potent melanotropic properties, had effects that were opposite from naturally occurring MSH on visual discrimination. This result was reminiscent of the opposing effect on avoidance conditioning of ACTH (4–10) (D-Phe-7) and the intact (4–10) sequence (DeWied, 1967), and underscores the high degree of behavioral specificity conferred by peptide structure. However, an Ala-1 and Lys-17 substitution in the ACTH (1–17) fragment produced a potent analog (Drago et al., 1984) that increased acquisition and delayed extinction of the active avoidance response. Adrenalectomy had no influence on this effect. Thus, some structural alterations may enhance the general "motivational" salience of stimuli, and this effect may be paradigm specific.

In this series of studies, only compounds with MSH-like conformations improved performance of the reversal learning problem. The initial learning phase may be activated nonspecifically, perhaps by general arousal mechanisms, whereas reversal learning may require specific peptide sequences interacting with their receptors in discrete areas of the brain. Thus, plural functions of similar peptides might result from their unique brain distribution and the cellular colocalization of peptides and transmitters. For instance, the arousal effects of the peptides may relate to patterns of colocalization in the reticular activating system. The attentional influence may result from another pattern of colocalization in another structure such as the hippocampus.

V. Behavioral Studies in Human Beings

A. NORMAL VOLUNTEERS

The effects of MSH/ACTH (4–10) have been reported to increase visual retention, decrease anxiety, and enhance visual discrimination (Miller et al., 1974; Sandman et al., 1975). Several processes have not been affected by the peptide, including short-term memory, measures of arousal, reaction time, and verbal memory. Although incompletely studied, inferential evidence indicated that the ACTH fragment influenced men and women differently (see Sandman et al., 1975; Veith et al., 1978). Several studies have explored processes affected by the peptide.

In the first study of perceptual processing, the influence of MSH/ACTH (4–10) on detection and discrimination was examined (Sandman et al., 1977a). Infusion of MSH/ACTH (4–10) impaired the subjects' ability to report accurately the presence of a stimulus presented for 6 msec. However, subjects' ability to discriminate accurately the two briefly presented stimuli was improved when given the peptide. These results were interpreted to suggest that MSH/ACTH (4–10) facilitated stimulus processing or selective attention but impaired the competing processes of simple detection. Conceivably the peptide changed the neuronal threshold for stimulus registration and functioned as a filtering mechanism to protect the organism from distracting "perceptual noise." When stimuli exceeded the threshold, processing of information was facilitated. This interpretation is consistent with the results of the event-related potential (ERP) studies discussed below (Rockstroh et al., 1981; Fehm-Wolfdorf et al., 1981; Sandman et al., 1986).

In a study (Ward et al., 1979) designed to test the influence of MSH/ACTH (4–10) on attention and memory, subjects were presented with a memory set consisting of 1, 2, 3, or 4 items. After they memorized the set, probe stimuli were presented. Half of the probes were members of the set and half were not. The subjects depressed one key if the probe was in the memorized set and a second key if it was not. Changes in memory are thought to be reflected by changes in the slope of reaction time plotted against set size. Improved memory (decreased slope) is illustrated by faster reaction time for set sizes 3 and 4 but not 1 and 2. Enhanced attention (encoding and response selection) may be inferred by faster reaction time at all set sizes and results in parallel functions for all experimental conditions. In this study, MSH/ACTH (4–10) resulted in a lower intercept but had no effect on the slope. In conjunction with other results, the most parsimonious interpretation of the altered intercept function is that MSH/ACTH (4–10) facilitated selective attention (encoding or response selection) to environmental stimuli.

B. MENTALLY RETARDED INDIVIDUALS

Gratifying effects of MSH/ACTH fragments on learning, attention, and memory have been observed in the behavior of mentally retarded individuals. Three studies have been completed. In the first study (Sandman et al., 1976), 20 mentally retarded men were injected with 15 mg of MSH/ACTH (4–10) and then given tests similar to those administered to normal volunteers. Treatment with the peptide resulted in a significant response to novel stimulation (an index of the orienting response), improved learning of intradimensional and extradimensional shifts, enhanced visual retention, and facilitated spatial localization and matching of auditory patterns.

In a second experiment (Walker and Sandman, 1979), the influence of the MSH/ACTH (4–9) analog was examined in a group of retarded adults. Three doses (0, 5, and 20 mg) were evaluated. The results indicated that, although significant improvement in measures of attention were observed (i.e., measures of concept learning requiring selective attention), the effects were neither as dramatic nor pervasive as in the initial study. Several factors may account for the attenuated effects, including reduced potency of the analog, oral route of administration, etc. However, a later study suggested that the choice of doses may have been the major reason.

In the third study, four doses (0, 5, 10, and 20 mg) of the ACTH (4–9) analog were examined in retarded clients while they performed their day-to-day activities requiring concentration, attention, and vigilance (Sandman et al., 1980b). The clients were paid a wage to bend electrical leads to fit a mold. There were four steps in the process which were graded according to difficulty. During the course of the study the clients performed the same task each day. The peptides were administered in the morning every day for two weeks. Measures of productivity and ratings of social behavior were made at regular intervals.

For the measures of productivity, the dose of the peptide interacted with the difficulty of the task. The high dose, 20 mg, interfered with the productivity for each level of difficulty. Mixed effects occurred with the 5-mg dose, enhancing performance only for the more complex tasks. Improved productivity in all but the easiest task occurred with 10 mg.

Administration of the peptide also influenced social behavior. Personal contact increased during treatment, especially with 10- and 20-mg doses. These results are in agreement with the animal studies discussed earlier, in which rats increased contact time after injections with the MSH and ACTH. In general, the studies indicated that treatment of retarded persons with ACTH fragment markedly improved their capacity. Whether this implies pituitary disregulation in this group (Sandman *et al.*, 1985) is uncertain. However, the possibility that ACTH-like peptides may be an adjunct for intervention is suggested.

C. ELDERLY SUBJECTS

A role for POMC and its ACTH-like fragments in the aging process has received mixed support. In a remarkable report, Landfield *et al.* (1981) found that the ACTH (4–9) analog arrested the aging process of rats. Using a number of aging mesures, these authors reported that adrenalectomy and ACTH (4–9) countered the effects of aging on reversal learning, neuronal density, nuclear "roundness," percentage of astrocytes, and an overall brain-aging index. They suggested that a "stimulant" influence of ACTH-like peptides (and a comparison drug, pentylenetetrazole) accounted for this effect.

An alternative, physiological explanation for the effects of neuropeptides on aging was proposed by DeWied and van Ree (1982). They reviewed the relative decline in a number of peptides in the brain with age. Clearly, the levels of the POMC fragments, MSH, ACTH, and β-endorphin, underwent considerable depletion with increasing age. Evidence was cited for the diminished levels of POMC and not its processing, suggesting that preprotein production and not enzymatic activity declined with age. Thus, as the bioavailability of POMC declines with age, neuronal pathways nurtured by its fragments may suffer degeneration.

There is equivocal evidence that MSH/ACTH fragments ameliorate the behavior of elderly human subjects (Ferris *et al.*, 1976, Branconnier *et al.*, 1979; Miller *et al.*, 1980). The study of Branconnier *et al.* is especially significant. Mildly senile, organically impaired subjects (18) of both sexes displayed reduced depression and confusion and increased vigor after treatment with MSH/ACTH (4–10). In addition, and consistent with reports by Gaillard and Varey (1979) and Born *et al.* (1984), the peptide delayed fatigue associated with a reaction time task. Surprisingly, the peptide produced a shift in the electroencephalogram (EEG) to lower frequencies (3.5–4.5 and 7.5–9.0 Hz). The authors indicated that these results were a function of responses in the women and suggested that the effects observed were evidence of a nonspecific arousing effect. By contrast, Miller *et al.* (1980) reported improvement in visual retention after MSH/ACTH (4–10) in the elderly, but the effect was greater in men than in women. Thus, there is some evidence of behavioral stimulation in the elderly after treatment with ACTH fragments, but a thorough study of these effects in humans has not been conducted.

VI. Electrophysiological Effects

A. ANIMALS

In a recent review, Urban (1984) has chronicled the effects of ACTH/MSH on the electrical activity of the brain. Using changes in brain electrical activity as an index of ACTH effects, he concluded that peripherally administered ACTH-like compounds influence the brain and that this influence may be a nonspecific change in neuronal responsivity to environmental stimulation. Urban reasoned that since the ACTH compounds facilitate adaptation, and since the organism cannot preconceive the nature of environmental challenge, the entire neuronal network may be sensitized.

Among the earliest reports of ACTH influences on the brain of rats were those indicating lowering of seizure threshold (Torda and Wolff, 1952), later reported for MSH (Izumi et al., 1973). In humans ACTH has been reported to ameliorate seizure activity (Gestaut et al., 1959), though more recent reports are less optimistic (Pentella et al., 1982; Hashimoto et al., 1981). However, in freely moving, conscious rats, injections of ACTH or MSH induced high amplitude, 4–9 Hz (slow) activity (Sandman et al., 1971). The significance of these findings has been explored more fully with computerized techniques in human subjects.

B. HUMAN SUBJECTS

In an early study of the influence of MSH/ACTH fragments, men received either MSH/ACTH (4–10) or ACTH (1–24) and were monitored for basal changes in physiological functions and during specific tasks (Miller et al., 1974). No effect on the EEG was observed in subjects receiving ACTH (1–24). However, spectral analysis of the EEG indicated that subjects injected with MSH/ACTH (4–10) had decreased power output of the 3–7 Hz frequency but increased power in the 8–12 and >12 Hz frequencies. The most striking finding in this study was the delay in the alpha-blocking EEG response to repetitive stimulation. Typically, during the first few trials there is a characteristic increase in EEG frequency to external stimulation. After several trials, the subject has habituated to the stimulus and the EEG response diminishes. Subjects treated with MSH/

ACTH (4–10) showed attenuation of the habituation and a persisting response.

The proliferation of computer averaging techniques has permitted very sophisticated analysis of brain-wave activity. Among the most useful of these techniques is averaging of the EEG during discrete periods of stimulation. The response observed is labeled an event-related potential (ERP) and in some way resembles a reflex of the brain. With many repetitions, a signal emerges that is linked to the stimulus. Components of the ERP are correlated with, and perhaps faithfully reflect, processes such as attention and memory. Responses occurring up to 200 msec after the stimulus typically reflect stimulus parameters (such as brightness or loudness) and the attention to, or perception of, them. Later components, 250–500 msec, are thought to measure organismic processes such as decision speed and memory. The ERP has been used to assess the effects of peptides.

In the initial study of somatosensory event-related potentials, administration of MSH increased the amplitude of early components of the first positive-going wave (P1) in both hypopituitary and normal patients (Kastin et al., 1971). In a later study of ERPs with the continuous performance task (CPT) as a stimulus, injections of MSH/ACTH (4–10) in normal volunteers resulted in increased latency and decreased amplitude of the second positive wave (P2) complex after visual stimulation (Miller et al., 1976). Even though these studies differed in several ways, changes in the early, stimulus-related components, rather than later, organismic components, were augmented by the MSH/ACTH peptide.

In a recent study of dosage and time effects (Sandman et al., 1986), early components again were influenced by the peptide. Men (5) and women (5) were given 0, 5, 10, and 20 mg of the analog or d-amphetamine (10 mg) as a positive control in a double-blind procedure. Immediately after ingesting the capsule, brief, bright flashes of light were projected, while EEGs were recorded from the right and left hemisphere of the occipital cortex. Of the orally administered ACTH (4–9) analog, 5 and 10 mg produced an initial effect on both power measures of the ERP and on P1 which peaked at 60 min. This effect followed the initial (30 min) suppression of P1. The effect of 20 mg followed the early time course seen with the 5- and 10-mg doses but failed to induce the "recovery" at 120 and 240 min. This effect was more evident in women than in men.

These results suggested that neural efficiency was enhanced by 60 min and persisted for several hours after administration of 5 or 10 mg of the ACTH (4–9) analog. No evidence of enhancement during this time period was apparent with the 20-mg dose. These results suggested either that the 5- to 10-mg dose is the optimal range for neural enhancement in human subjects or that the dosage interaction with time has not been completely explored. Thus, the 20-mg dose may show "recovery" (and, therefore, neural/behavioral enhancement) beyond the 240 min studied. Further studies are required to explore these possibilities.

A recent collection of studies has examined the effect of ACTH fragments on early and late components of ERP (Fehm-Wolfsdorf *et al.*, 1981; Rockstroh *et al.*, 1981, 1983; Born *et al.*, 1984). The results, in general, supported the previous findings that ACTH fragments primarily influence the early components. However, under some conditions, later components also changed after peptide administration. Rockstroh *et al.* (1981) administered 40 mg of the ACTH (4–9) analog and examined its effect on simple reaction time (RT) and the ERP. Both the latency of RT and of the first negative-going wave (N1) decreased in association with peptide treatment. Although they reported increased amplitude of early and late components (N1 and P3), they failed to reach acceptable levels of statistical significance. In a variation of this procedure, Fehm-Wolfsdorf *et al.* (1981) presented a warning stimulus, either a high or a low tone, then a second stimulus that was either neutral or aversive. Treatment with the ACTH compounds resulted in a persistent decrement in RT to both stimuli. The P3 response was small in the presence of the ACTH analog, an effect that was different from the Rockstroh *et al.* (1981) finding.

In another variation (Rockstroh *et al.*, 1983), subjects were warned and then either distracted or not distracted before a second stimulus was presented. The response to the second stimulus was assessed after treatment with the ACTH analog. Subjects given the peptide had larger N1 responses to the second stimulus in the distraction phase; however, RT was not affected. The authors speculated, consistent with earlier observations (Sandman *et al.*, 1977), that the ACTH analog may have promoted a threshold effect serving to filter environmental input. As such, and as proposed by Urban (1984), ACTH may sensitize neuronal networks in restricted areas of the brain for enhancement of adaptive responses to specific environmental challenge.

VII. Developmental Studies: Organizational Influences of Neuropeptides on the Brain

The studies discussed up to this point have all been conducted with mature, adult animals. In each case the effects described, although often dramatic, persisted for only a short time, perhaps hours, without any apparent lingering consequences. These effects are activational. The following studies describe the effects of peptides of the POMC molecule on the brain and behavior of the immature nervous system. Since the influence of early exposure of peptides can persist for the lifetime of the organism, these effects may be viewed as organizational.

Phoenix *et al.* (1959) were the first to propose organizational endocrine influences on behavior by observing the masculinizing effects of prenatally administered testosterone on adult female guinea pigs. They reasoned that testosterone, rather than the individual's genetic sex, had a masculinizing effect on the soma during critical perinatal periods. Firm support for the organizational–activational role of the hypothalamic–pituitary–gonadal axis has been obtained across a wide variety of species.

Early interventions of the hypothalamic–pituitary–thyroid system have a radical impact on later capacity to interact with the environment. Neonatal rats undergoing thyroidectomy have severe retardation of physiological, reflex, and central nervous system (CNS) functioning that persist into adulthood (Eayrs, 1961; Eayrs and Levine, 1963). In contrast, neonatal activation of the hypothalamic–pituitary–thyroid axis during early periods may enhance aspects of the rats' later development (Stratton *et al.,* 1976).

The effects of early manipulation of the hypothalamic–pituitary–adrenocortical axis by early handling resulted in a greater adaptive, or "economical," functioning of that system in adulthood. Rats handled in infancy exhibited a more rapid and intense, yet less prolonged, corticosteroid response to shock in adulthood when compared with nonhandled animals (Bell *et al.,* 1961; Levine, 1962). Such treatment has also been found to decrease responsivity to novel open-field situations, as indicated by lowered corticosterone secretion (Levine *et al.,* cited in Levine and Mullins, 1966).

ACTH and especially MSH have a major role during early development since they stimulate fetal growth and brain development in the immature organism (Swaab *et al.,* 1978). Maternal blood levels of MSH are related to the onset of labor, rise during the initiation of labor, remain elevated until birth, and then return to normal levels (Clark *et al.,* 1978). It is conceivable that the fetus is exposed to high levels of MSH just prior to birth, and variations may have a lasting influence on behavior and the brain. In two recent studies, the effect of elevated ACTH on the fetus was explored. In one report (Monder *et al.,* 1980), ACTH given 5 days pre- and postpartum resulted in growth retardation, delayed eye-opening, and restricted vaginal opening. These effects were blocked by the simultaneous administration of naloxone. In a second study (Stylianopoulou, 1983) rats were exposed to ACTH during the last half of pregnancy. The result indicated that females were masculinized and the males demasculinized. This effect, however, was only apparent in some litters.

Slightly different results were obtained with neonatal exposure. Neonatal (days 2–7) administration of MSH to rats increased later performance of several tasks (Beckwith *et al.*, 1977b). In an operant DRL-20 task, in which hungry juvenile rats had to learn to withhold a barpress response at least 20 sec to receive food, neonatal peptide treatment resulted in significantly increased efficiency. In the same report (but with different animals), early exposure to MSH improved avoidance and extinction learning in rats. Early exposure to either MSH or the ACTH (4–9) analog (Champney *et al.*, 1976) improved visual discrimination and reversal learning. In this later study, ICV injections had effects identical to peripheral injections. Furthermore, early injections sensitized animals to later exposure to the ACTH (4–9) analog.

It is clear that perinatal exposure of rats to peptides from the ACTH family exerts persisting effects on the brain and behavior. Although in the studies reviewed it is difficult to isolate the direct effects of dosage, it would appear that fetal exposure to ACTH might induce teratogenic effects, whereas neonatal administration enhances adaptability. This may be an oversimplified analysis for several reasons, including the possibility that MSH and not ACTH sequences are physiologically relevant for the fetus (Swaab *et al.*, 1978). In any case, the results of these studies extend the number of endogenous chemicals that assist in the organization of neural processes.

VIII. Endogenous Levels

Another strategy for assessing the effects of ACTH-like peptides is examination of the relationship between levels of the peptide and ongoing behavior. The widely used dexamethasone suppression test for the diagnosis of depression implicates the role of ACTH, or at least the integrity of pituitary–adrenal axis, in the cognitive/affective state of some patients. However, few direct studies of the relationship between plasma or brain levels of ACTH and behavior exist.

Circulating levels of endogenous ACTH have been reported to influence sensory thresholds. Henkin (1975), in his review, reported that chronic adrenocortical insufficiency (resulting in high levels of ACTH) significantly enhanced abilities to detect gustatory, olfactory, and auditory stimuli. Correlational studies have demonstrated that rats with the highest plasma corticosteroid response to ether-induced stress also exhibited the greatest proficiency in acquiring the correct response in an avoidance paradigm (Wertheim *et al.*, 1969). The rate of acquisition of the avoidance response also has been associated with circadian fluctuations of ACTH with optimal performance observed during the peak of the ACTH cycle (Pagano and Lovely, 1972; Schneider et al., 1974). However, chronically elevated levels of 17-hydroxycorticosterone (17-OHCS) in rhesus monkeys have been associated with lower acquisition rate of operant avoidance responding (Levine et al., 1970). Removal of pituitary ACTH by hypophysectomy retarded acquisition and facilitated extinction in both active and passive avoidance situations (Applezweig and Baudry, 1955; Weiss et al., 1970). Adrenalectomy induced chronic high elevations of ACTH and facilitated avoidance behavior (Beatty et al., 1970; Weiss et al., 1970).

A recent study in humans (Veith et al., 1985) carefully examined the relationship between ACTH and behavior. Patients ($N = 8$) with congenital adrenal hyperplasia (CAH) were examined. CAH is an autosomal genetic defect with several consequences including severe cortisol deficiency. The cortisol deficiency triggers a marked elevation of ACTH. These patients received cortisone medication to supress ACTH release. However, during this study, the levels of ACTH (which were measured) were manipulated by controlling the medication. Thus, ACTH levels were either suppressed or elevated. During periods of elevated ACTH, the patients had faster RTs on all sets of the item recognition test (an effect of intercept, not slope). Indeed the relationship between plasma ACTH levels and the intercept (RT) was -0.68 (df = 10, $p < 0.05$). These findings are remarkably consistent with those of Ward et al. (1979) in which normal volunteers were administered fragments of ACTH. Of interest in the study by Veith et al. is the cognitive enhancement despite the withdrawal of needed medication. Even as the physiological system was allowed to "free run," seemingly disrupted, there was a benefit of enhanced performance because, ostensibly, of the elevated ACTH.

IX. Conclusion

ACTH/MSH and the family of related molecules exert a number of behavioral influences including (1) the induction of grooming, (2) stretching and yawning, (3) normalization of social behavior, (4) opiate-like effects, (5) improved learning, attention, and memory, (6) characteristic effects on the computerized EEG, and (7) organizational effects on the developing nervous system. Furthermore, there are encouraging clinical reports of the effects of ACTH fragments in the mentally retarded and the elderly. This panorama of effects defies a simple, singular summary. The stress-related theory of ACTH effects certainly can be supported. Our earlier evidence (Sandman et al., 1977b, 1981, 1982) that part of the

ACTH molecule improved attention also can be supported. One apparent paradox is the mutual exclusivity of some of these conclusions. Perhaps a more catholic explanation is required.

An explanation that incorporates the findings of Strand and Smith (1980) and Saint-Come et al. (1982) would need to be broader. These researchers have found that some, but not all, fragments of ACTH had beneficial effects on the "functional reorganization of regenerating motor units" of the extensor digitorium longus after crushing of the peroneal nerve. Both motor unit activity and neuromuscular efficiency increased after administration of ACTH. Strand's view is that ACTH increases neuronal plasticity. She supports this view with observations that ACTH increased protein synthesis in nerves, promoted rapid growth of endplates, enhanced regeneration of axonal fibers, and produced proliferation of preterminal nerve fibers.

The exciting report of Flohr and Luneburg (1982) is very consistent with Strand's finding. These authors performed unilateral labyrinthectomy on *Rana temporaria* and examined the compensatory processes after treatment with MSH/ACTH (4–10). If treatment began immediately at high doses (250 µg/day), compensation and its maintenance was very significantly enhanced. If treatment was withdrawn, there was a gradual decline to the level of untreated controls. The authors speculated that compensation progressed in an orderly sequence that was goal-directed, adaptive, and in some ways resembled learning. They posited that ACTH promoted plasticity by rectifying activity to optimize adaptation.

The recent report of Long and Haladay (1985) that ACTH increased the permeability of the brain–blood barrier may provide a mechanism for the actions of ACTH. Rather than suggest that ACTH has the range of effects we have described, including these remarkable reports of increased plasticity, perhaps ACTH simply made neuronal systems more accessible to blood-borne materials. In this regard, MSH was the first peptide for which chromatographic evidence of penetration of the blood–brain barrier was provided (Kastin et al., 1976).

The brain as a target organ remains an appealing hypothesis and may be the only mechanism that explains some of the unique, sequence-specific behavioral pattern reported. Conceivably, both a general (central) and specific (peripheral) mechanism for the influence of ACTH may coexist. At this point both views can be supported.

References

Applezweig, M. H., and Baudry, F. D. (1955). *Psychol. Rep.* **1**, 417–420.
Beatty, P. A., Beatty, W. W., Bowman, R. E., and Tilchrist, J. C. (1970). *Physiol. Behav.* **5**, 939–944.

Beckwith, B. E., O'Quin, R. K., Petro, M. S., Kastin, A. J., and Sandman, C. A. (1977a). *Physiol. Psychol.* **5**, 295–299.
Beckwith, B. E., Sandman, C. A., Hothersall, D., and Kastin, A. J. (1977b). *Physiol. Behav.* **18**, 63–71.
Bell, R. W., Reisner, G., and Linn, T. (1961). *Science* **133**, 1428.
Bertolini, A., and Gessa, G. L. (1981). *J. Endocrinol. Invest.* **4**, 241–257.
Bertolini, A., and Gessa, G. L., and Ferrari, W. (1975). In "Sexual Behavior Pharmacology and Biochemistry" (M. Sandler and G. L. Gessa, eds.), p. 247. Raven, New York.
Bertolini, A., Poggioli, R., and Ferrari, W. (1979). *Experientia* **35**, 1216.
Bertolini, A., Poggioli, R., and Ferrari, W. (1980). In "Neural Peptides and Neural Communication" (E. Costa and M. Trabucchi, eds.), pp. 109–117. Raven, New York.
Born, J., Fehm-Wolsdorf, G., Schieke, M., Rockstroh, B., Fehm, H. C., and Voight, K. H. (1984). *Pharmacol. Biochem. Behav.* **21**, 513–519.
Borvendez, J., Graf, L., Hermann, I., Palkovits, M., and Meretey, K. (1978). In "Endorphins' 78" (L. Graf, M. Palkovits, and A. Z. Ronai, eds.), pp. 182–187. Akademiai Kiado, Budapest.
Branconnier, R. J., Cole, J. O., and Gardos, G. (1979). *Psychopharmacology* **61**, 161–165.
Britton, D. R., Hoffman, D. K., Lederij, K., and Rivier, J. (1984). *Brain Res.* **304**, 201–205.
Champney, T. F., Shaley, T. C., and Sandman, C. A. (1976). *Pharmacol. Biochem. Behav.* **5**, 3–10.
Civelli, O., Birnberg, N., and Herbert, E. (1982). *J. Biol. Chem.* **297**, 6783–6787.
Clark, D., Thody, A. J., Shuster, S., and Bowers, H. (1978). *Nature (London)* **273**, 163–164.
Crabbe, J. C., Rigter, H., and Herbusch, S. (1982). *Behav. Brain Res.* **4**, 289–314.
Crawley, J. N., Hays, S. E., O'Donohue, T. L., Paul, S. M., and Goodwin, F. K. (1981). *Peptides* **2** (Suppl.1), 123–129.
de Almeida, M. A. M. R., Kapczinski, F. P., and Izquierdo, I. (1983). *Behav. Neural Biol.* **39**, 272–283.
DeWied, D. (1964). *Am. J. Physiol.* **207**, 255–259.
DeWied, D. (1967). *Excerpta Med. Int. Congr. Ser.* **132**, 945.
DeWied, D. (1983). *Acta Morphol. Hung.* **31**, 1–3.
DeWied, D., and Bohus, B. (1966). *Nature (London)* **212**, 1484–1488.
DeWied, D., and Jolles, J. (1982). *Physiol. Rev.* **62**, 976–1059.
DeWied, D., and van Ree, J. M. (1982). *Life Sci.* **31**, 709–719.
DeWied, D., Bohus, B., Van Ree, J. M., and Urban, I. (1978). *J. Pharmacol. Exp. Ther.* **204**, 570–580.
Dorsa, D. M., de Kloet, E. R., Van Dijk, A. M. J., and Mezey, E. (1982). In "Neurosecretion: Molecules, Cells and Systems" (K. Lederis and D. Farner, eds.), pp. 199–211. Plenum, New York.
Drago, F., Continella, G., and Scapagnini, U. (1984). *Pharmacol. Biochem. Behav.* **20**, 689–695.
Dunn, A. J., and Jurd, R. W. (1984). *Brain Res. Bull.* **12**, 369–371.
Eayrs, J. T. (1961). *J. Endocrinol.* **22**, 409–419.
Eayrs, J. T., and Levine, S. (1963). *J. Endocrinol.* **25**, 505–515.
Eskay, R. L., Giravd, P., Oliver, C., and Brownstem, M. J. (1979). *Brain Res.* **178**, 55–67.
Evans, C. J., Erdelyi, E., Weber, E., and Barchas, J. D. (1983). *Science* **221**, 957–960.
Fehm-Wolsdorf, G., Elbert, T., Lutzenberger, W., Rochstroh, B., Birbaumer, N., and Fehm, H. C. (1981). *Psychoneuroendocrinology* **6**, 311–320.
Ferrari, W., Floris, E., and Paulesu, F. (1955). *Boll. Soc. It. Biol. Sper.* **31**, 862.
Ferrari, W., Gessa, G. L., and Vargiu, L. (1963). *Ann. N.Y. Acad. Sci.* **104**, 330.
Ferris, S. H., Sathananthan, G., Gershon, S., Clark, C., and Moshinsky, J. (1976). *Pharmacol. Biochem. Behav.* **5**, 73–78.

Flohr, H., and Luneburg, U. (1982). *Brain Res.* **248**, 169–173.
Gaillard, A. W. K., and Varey, C. A. (1979). *Physiol. Behav.* **23**, 79–84.
Genedani, S., Bernardi, M., and Bertolini, A. (1984). *Neuropeptides* **4**, 247–250.
Gessa, G. L., Pisano, M., Vargiu, L., Crabai, F., and Ferrari, W. (1967). *Rev. Can. Biol.* **26**, 229.
Gestaut, H., Miribel, G., Flavel, P., and Vigoroux, M. (1959). *Soc. Fra. Neurol.* **101**, 753–762.
Gispen, L. H. (1982). *Acta Biol. Med. Germ.* **41**, 279–288.
Gispen, W. H., and Isaacson, R. L. (1981). *Pharmacol. Ther.* **12**, 209–246.
Gispen, W. H., Buitelaar, J., Wiegnant, V. N., Trenius, L., and de Wied, D. (1976). *Eur. J. Pharmacol.* **39**, 393.
Gold, P. E., and van Buskirk, R. (1976a). *Behav. Biol.* **16**, 387–400.
Gold, P. E., and van Buskirk, R. (1976b). *Horm. Behav.* **7**, 509–517.
Gramsch, C., Kleber, G., Hollt, V., Pasi, A., Mehraein, P., and Herz, A. (1980). *Brain Res.* **192**, 109–119.
Greven, H. M., and DeWied, D. (1977). *Front. Horm. Res.* **4**, 140–148.
Hashimoto, T., Hivra, K., Suzue, J., Kokawa, T., Fukuda, K., Endo, S., Tayama, M., Tamvra, Y., and Miyao, M. (1981). *Brain Dev.* **3**, 51–56.
Henkin, R. I. (1975). *In* "Anatomical Neuroendocrinology" (W. E. Stumpf and L. D. Grant, eds.). Karger, Basal.
Izumi, K., Donaldson, J., Minnich, J., and Barbeau, A. (1973). *Can. J. Physiol. Pharmacol.* **51**, 572–578.
Jacquet, Y. F., and Wolf, G. (1981). *Brain Res.* **219**, 214–218.
Jolles, J., Rompa-Barendregt, J., and Gispen, J. (1979). *Horm. Behav.* **12**, 60.
Kastin, A. J., Miller, L. H., Gonzalez-Barcena, D., Hawley, W. D., Dyster-Aas, K., Schally, A. V., Parra, M. L. V., and Velasco, M. (1971). *Physiol. Behav.* **7**, 893–896.
Kastin, A. J., Nissen, C., Nikolics, K., Medzihradszky, K., Coy, D. H., Teplan, I., and Schally, A. V. (1976). *Brain Res. Bull.* **1**, 19–26.
Kastin, A. J., Zadina, J. E., Coy, D. H., Schally, A. V., and Sandman, C. A. (1980). *In* "Polypeptide Hormones" (R. F. Beers and E. G. Bassett, eds.), pp. 223–233. Raven, New York.
Kastin, A. J., Olson, R. D., Sandman, C. A., and Coy, D. H. (1981). *In* "Endogenous Peptides and Learning and Memory Processes" (J. L. Martinez, Jr., R. A. Jensen, R. B. Messing, H. Rigter, and J. L. McGaugh, eds.), pp. 563–577. Academic Press, New York.
Kastin, A. J., Banks, W. A., Zadina, J. G., and Graf, M. (1983). *Life Sci.* **32**, 295–301.
Kastin, A. J., Zadina, J. E., Banks, W. A., and Graf, M. (1984). *Peptides (Suppl.)* **1**, 249–253.
Kobobun, K., O'Donohue, T. C., Handelmann, G. D., Sawyer, T. K., Hruby, V. J., and Hadley, M. E. (1983). *Peptides* **4**, 721–724.
Krieger, D. T., Liotta, A., Suda, T. *et al.* (1977). *Biochem. Biophys. Res. Commun.* **76**, 930–936.
Krieger, D. T., Liotta, A. S., Brownstein, M. J., and Zimmerman, E. A. (1980). *Recent Prog. Horm. Res.* **36**, 277–338.
Landfield, P. W., Baskin, R. K., and Pitler, T. A. (1981). *Science* **214**, 581–583.
Levine, M. D., Gordon, T. P., Peterson, R. H., and Rose, R. M. (1970). *Physiol. Behav.* **5**, 919–924.
Levine, S. (1962). *Science* **135**, 795–796.
Levine, S., and Mullins, R. F., Jr. (1966). *Science* **152**, 1585–1592.
Long, J. B., and Haladay, J. W. (1985). *Science* **227**, 1580–1583.

McGaugh, J. L. (1983). *Am. Psychologist* **38**, 161–174.
MacKintosh, N. J. (1965). *Psychol. Bull.* **64**, 124–150.
MacKintosh, N. J. (1969). *J. Comp. Physiol. Psychol.* **67**, 1–18.
Mason, J. W. (1975). *J. Hum. Stress* **1**, 22–36.
Mezey, E. M., and Palkovits, M. (1982). *Front. Neuroendocrinol.* **7**, 1–29.
Miller, L. H., Kastin, A. J., Sandman, C. A., Fink, M., and van Veen, W. J. (1974). *Pharmacol. Biochem. Behav.* **2**, 663–668.
Miller, L. H., Harris, L. C., Van Riezen, H., and Kastin, A. J. (1976). *Pharmacol. Biochem. Behav.* **5**, 17–22.
Miller, L. H., Groves, G. A., Bupp, M. J., and Kastin, A. J. (1980). *Peptides* **1**, 55–57.
Miller, R. E., and Ogawa, N. (1962). *J. Comp. Physiol. Psychol.* **55**, 211–213.
Mirsky, I. A., Miller, R., and Stein, M. (1953). *Psychosom. Med.* **15**, 574–588.
Moldow, R. L., and Yalow, R. S. (1978). *Proc. Natl. Acad. Sci. U.S.A.* **75**, 994–998.
Monder, H., Yasukawa, N., and Christain, J. J. (1980). *Horm. Behav.* **14**, 329–336.
Mousa, S., and Couri, D. (1983). *Substance Alcohol Act./Misuse* **4**, 1–18.
Murphy, J. V., and Miller, R. E. (1955). *J. Comp. Physiol. Psychol.* **48**, 47–49.
Niesink, R. J. M., and van Ree, J. M. (1983). *Science* **221**, 960–962.
Niesink, R. J. M. and van Ree, J. M. (1984). *Neuropeptides* **4**, 483–496.
O'Donohue, T. C., Handelmann, G. E., Loh, Y. P., Olton, D. S., Lizbowitz, J., and Jacobowitz, D. M. (1981a). *Peptides* **2**, 101–104.
O'Donohue, T. C., Handelmann, G. E., Chaconas, T., Miller, R. L., and Jacobowitz, D. M. (1981b). *Peptides* **2**, 333–344.
O'Donohue, T. L., Handelmann, G. E., Miller, R. L., and Jacobowitz, D. M. (1982). *Science* **215**, 1125–1127.
Pagano, R. R., and Lovely, R. H. (1972). *Physiol. Behav.* **8**, 721–723.
Pentella, K., Bachman, D. S., and Sandman, C. A. (1982). *Neuropediatrics* **13**, 59–62.
Phoenix, C. H., Goy, R. W., Gerall, A. A., and Young, W. D. (1959). *Endocrinology* **65**, 369–382.
Piekut, D. T., and Knigge, K. M. (1984). *Peptides* **5**, 1089–1095.
Rigter, H. (1978). *Science* **200**, 83–85.
Rigter, H., and van Riezen, H. (1975). *Physiol. Behav.* **14**, 563–566.
Rigter, H., Jamssens-Elbertse, R., and van Riesen, H. (1976). *Pharmacol. Biochem. Behav.* **5**, (Suppl. 1), 53–58.
Rockstroh, B., Elbert, T., Lutzenberger, W., Birbaumer, N., Fehm, H. C., and Voight, K. H. (1981). *PNE* **6**, 301–310.
Rockstroh, B., Elbert, T., Lutzenberger, W., Birbaumer, N., Boight, K. H., and Fehm, H. C. (1983). *Int. J. Neurosci.* **22**, 21–36.
Rudman, D., Scott, J. W., Del Rio, A. E., Houser, H., and Sheen, S. (1974). *Am. J. Physiol.* **226**, 682–686.
Saint-Come, C., Acker, G. R., and Strano, F. L. (1982). *Peptides* **3**, 439–449.
Sandman, C. A., and Kastin, A. J. (1981a). *Pharmacol. Ther.* **13**, 39–60.
Sandman, C. A., and Kastin, A. J. (1981b). *Peptides* **2**, 231–133.
Sandman, C. A., and O'Halloran, J. P. (1986). *In* "Encyclopedia on Pharmacology and Therapeutics" (D. Dewied, W. H. Gispen, and Tj.B. van Wimersma Greidanus, eds.), pp. 397–420. Pergamon, Oxford.
Sandman, C. A., Denman, P., Miller, L. H., Knott, J. R., Kastin, A. J., and Schally, A. V. (1971). *J. Comp. Physiol. Psychol.* **76**, 303–310.
Sandman, C. A., Miller, L. H., Kastin, A. J., and Schally, A. V. (1972). *J. Comp. Physiol. Psychol.* **80**, 54–58.
Sandman, C. A., Alexander, W. D., and Kastin, A. J. (1973). *Physiol. Behav.* **11**, 613–617.

Sandman, C. A., Beckwith, B. E., Gittis, M. M., and Kastin, A. J. (1974). *Physiol. Behav.* **13**, 163–166.
Sandman, C. A., George, J., Nolan, J. D., Van Riezen, H., and Kastin, A. J. (1975). *Physiol. Behav.* **15**, 427–431.
Sandman, C. A., George, J., Walker, B., Nolan, J. D., and Kastin, A. J. (1976). *Pharmacol. Biochem. Behav.* **5**, 23–28.
Sandman, C. A., George, J., McCanne, T. R., Nolan, J. D., Kaswan, J., and Kastin, A. J. (1977a). *J. Clin. Endocrinol. Metab.* **44**, 884–891.
Sandman, C. A., Kastin, A. J., and Miller, L. H. (1977b). *In* "Clinical Neuroendocrinology" (L. Martini and G. M. Besser, eds.), pp. 443–470. Academic Press, New York.
Sandman, C. A., Beckwith, B. E., and Kastin, A. J. (1980a). *Peptides* **1**, 277–280.
Sandman, C. A., Walker, B. B., and Lawton, C. A. (1980b). *Peptides* **1**, 109–114.
Sandman, C. A., Kastin, A. J., and Schally, A. V. (1981). *In* "Neuroendocrine Regulation and Altered Behavior" (P. S. Hrdina and R. L. Singhal, eds.), pp. 5–27. Croom Helm, London.
Sandman, C. A., Barron, J., and Parker, L. (1985). *Pharmacol. Biochem. Behav.* **23**, 21–26.
Sandman, C. A., Berka, C., Veith, J. L., and Walker, B. B. (1985). *Peptides* 803–807.
Schneider, A. M., Weinberg, J., and Weissberg, R. (1974). *Physiol. Behav.* **13**, 633–636.
Selye, H. (1936). *Nature (London)* **138**, 32–33.
Selye, H. (1956). "The Stress of Life." McGraw-Hill, New York.
Strand, F. L., and Smith, C. M. (1980). *Pharmacol. Ther.* **11**, 509–533.
Stratton, L. O., and Kastin, A. J. (1974). *Horm. Behav.* **5**, 149–155.
Stratton, L. O., Gibson, C. A., Kolar, K. G., and Kastin, A. J. (1976). *Pharmacol. Biochem. Behav.* **5** (Suppl. 1), 65–67.
Stylianopoulou, F. (1983). *Horm. Behav.* **17**, 324–331.
Swaab, D. F., Boer, G. J., Boer, K., Dogterom, J., van Leevwen, F. W., and Visser, M. (1978). *In* "Maturation of the Nervous System, Progress in Brain Research" (M. A. Corner, R. E. Baker, N. E. van de Poll, D. F. Swabb, and H. B. M. Uylings, eds.). Elsevier, Amsterdam.
Thody, A. J., and Wilson, C. A. (1983). *Physiol. Behav.* **31**, 67–72.
Tinklenberg, J. R., and Thorton, J. D. (1983). *Psychopharmacol. Bull.* **19**, 198–211.
Torda, C., and Wolff, H. G. (1952). *Am. J. Physiol.* **168**, 906–913.
Urban, I. J. A. (1984). *Pharmacol. Ther.* **24**, 57–90.
Van Ree, J. M., Bohus, B., Csontos, K., Gispen, W. H., Gaeven, H. M., Nijkame, F. P., Dpmer, F. A., de Rotte, G. A., van Wimersma Greidanus, T. B., Witter, A., and DeWied, D. (1981). *Life Sci.* **28**, 2875–2888.
van Wimersma Greidanus, T. B., Dijk, A. M. A., van de Rotte, A. A., Goedemans, J. H. J., Croiset, G., and Thody, A. J. (1978). *Brain Res. Bull.* **3**, 227–230.
Vecesi, L., Teleody, G., Schally, A. V., and Coy, D. H. (1981). *Peptides* **3**, 398–391.
Veith, J. L., and Sandman, C. A. (1985). *In* "Physiological and Psychological Interactions" (S. R. Burchfield, ed.), pp. 129–161. Hemisphere, Washington, D.C.
Veith, J. L., Sandman, C. A., George, J., and Stevens, V. C. (1978). *Physiol. Behav.* **20**, 43–50.
Veith, J. L., Sandman, C. A., George, J. M., and Kendall, J. W. (1985). *Psychoneuroendocrinology* **10**, 33–48.
Walker, B. B., and Sandman, C. A. (1979). *Am. J. Ment. Defic.* **83**, 346–352.
Walker, J. M., Brentson, G. B., Sandman, C. A., Kastin, A. J., and Akil, H. (1981). *Eur. J. Pharmacol.* **69**, 71–79.
Ward, M. M., Sandman, C. A., George, J., and Shulman, H. (1979). *Physiol. Behav.* **22**, 669–673.

Weiss, J. M., McEwen, B. S., Silva, M. T., and Kalkut, M. (1970). *Am. J. Physiol.* **218,** 864–868.
Wertheim, G. A., Conner, R. L., and Levine, S. (1969). *Physiol. Behav.* **4,** 41–44.
William, N. S., and Scampoli, D. L. (1984). *Pharmacol. Biochem. Behav.* **20,** 681–682.
Winter, Ch. A., and Flataker, L. (1951). *J. Pharmacol. Exp. Ther.* **101,** 93.

7
Regulation of ACTH Secretion and Synthesis

TERRY D. REISINE AND JULIUS AXELROD

Laboratory of Cell Biology
National Institute of Mental Health
Bethesda, Maryland 20205

I. Introduction

The body reacts to some environmental and physiological stimuli by secreting adrenocorticotropin (ACTH) from the anterior pituitary (Axelrod and Reisine, 1984). Many investigators observed that a variety of stressful events causes the release of ACTH (Yates and Maran, 1974). Harris (1948) demonstrated that ACTH release from the pituitary is regulated by a corticotropin releasing factor (CRF) synthesized in the hypothalamus. After a long period of intensive investigations, CRF was isolated, purified, and its structure characterized as a 41-amino acid peptide by Vale *et al.* (1981) and co-workers (Spiess *et al.*, 1981). A number of *in vivo* and *in vitro* studies has since shown that CRF is the most potent and effective natural stimulant of ACTH secretion.

The availability of synthetic CRF and of a mouse anterior pituitary cell line (AtT-20/D16-16) that secretes ACTH made it possible to study the intracellular mechanisms involved in the release of ACTH (Axelrod and Reisine, 1984). Modern approaches to investigate ACTH secretion from the pituitary have mainly employed primary cultures of the rat adenohypophysis (Labrie *et al.*, 1982; Giguere *et al.*, 1982). Although much useful information has been obtained with such a preparation, the heterogeneity of the cell types and the low density of the ACTH-secreting cells (2–3% of the total cell population) in the anterior pituitary have limited the characterization of factors directly controlling ACTH release. The AtT-20 cell line has been used previously to examine the processing of the precursor protein proopiomelanocortin (POMC) to generate ACTH as well as the storage and secretion of ACTH and β-endorphin (Mains and Eipper, 1976; Roberts et al., 1978; Sabol, 1980). This anterior pituitary cell line can be propagated and appears to be homogeneous with regard to cell type, and, in contrast to primary anterior pituitary cultures, these cells predominantly release hormones of the POMC family of peptides.

Normal anterior pituitary cells respond to synthetic CRF by releasing ACTH and β-endorphin (Vale et al., 1981; Spiess et al., 1981). AtT-20 cells were also found to secrete immunoreactive ACTH and β-endorphin in response to CRF (Hook et al., 1982). Analogs of CRF show the same order of potency in releasing ACTH from AtT-20 cells as observed in normal corticotrophs, and, as shown with intact animals and primary cultures of the pituitary, glucocorticoids blocked the CRF-stimulated release of ACTH from the tumor cells (Hook et al., 1982). These findings prompted the use of AtT-20 cells as a model for investigating the cellular and molecular mechanisms that regulate ACTH secretion from the anterior pituitary.

II. Multireceptor Release of ACTH

A. Catecholamines

Beside CRF, other hormones stimulate ACTH release from the anterior pituitary (Vale and Rivier, 1977; Berkenbosch et al., 1981; Tilders et al., 1982) (Fig. 1). Norepinephrine was shown to evoke ACTH release from AtT-20 cells (Mains and Eipper, 1981), and ligand binding studies using tritiated dihydroalprenolol, a β-adrenoceptor antagonist, indicated the presence of a β-adrenoceptor on AtT-20 cells (Reisine et al., 1983). Isoproterenol, a β-adrenoceptor agonist, as well as epinephrine induced a potent and stereoselective increase of ACTH release from mouse tumor cells which was calcium-dependent and blocked by the β-adrenoceptor antagonist, propranolol (Reisine et al., 1983). Two subtypes of β-adrenoceptors are known, with β_2-adrenoceptors being most sensitive to epinephrine while β_1-adrenoceptors are equally responsive to epinephrine and norepinephrine (Furchgott, 1972). Pharmacological characterization showed that β_2-receptors are present on AtT-20 cells and could mediate the release of ACTH. The presence of β_2- but not β_1-adrenoceptors has also been reported in the rat anterior pituitary (Petrovic et al., 1983).

B. Vasoactive Intestinal Peptide

Vasoactive intestinal peptide (VIP), present in the hypothalamus and known to stimulate prolactin release from the anterior pituitary (Rotsztejn

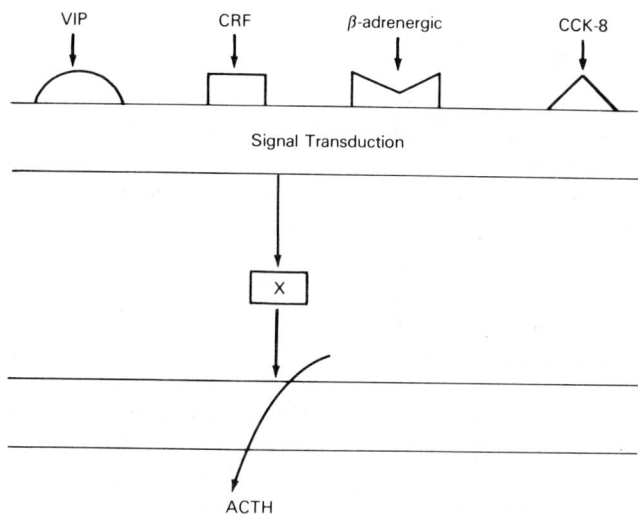

Fig. 1. Multihormonal control of ACTH release. The secretion of ACTH from AtT-20 cells is stimulated by vasoactive intestinal peptide (VIP), corticotropin releasing factor (CRF), catecholamines acting upon β-adrenergic receptors, and cholecystokinin-8 (CCK-8). Each hormone stimulates an independent receptor to evoke ACTH release. The precise cascade of intracellular events initiated by each hormone to release ACTH is unknown (X).

et al., 1980), also evokes the secretion of ACTH from AtT-20 cells in a dose-dependent manner (Reisine *et al.*, 1982). VIP was also observed to stimulate ACTH release from human anterior pituitary tumor cells (Olivia *et al.*, 1982) and from rat anterior pituitary primary cultures at high concentrations (Westendorf *et al.*, 1983). The effect of VIP on ACTH secretion, like that of other hormones, is blocked by glucocorticoids (Reisine *et al.*, 1982).

C. Vasopressin

Arginine-vasopressin is also found in hypothalamic neurons and was one of the first hormones proposed to have CRF-like actions. Vasopressin increases the secretion of ACTH from primary cultures of the anterior pituitary (Giguere and Labrie, 1982; Vale *et al.*, 1983). However, its major action appears to be a potentiation of the ACTH release response to CRF (Yates *et al.*, 1971). This result is of interest since CRF and vasopressin have been found to be colocalized in some of the same paraventricular neurons of the hypothalamus that project to the median eminence (Sawchencko *et al.*, 1984; Kiss *et al.*, 1984). This potentiating action of vasopressin may therefore serve as a physiological function of neurotransmitter colocalization.

D. Cholecystokinin

Another hypothalamic peptide with putative ACTH releasing properties is cholecystokinin-8 (CCK8). This substance stimulates ACTH secretion from both primary cultures of the rat anterior pituitary and AtT-20 cells (Reisine and Jensen, 1986). Neither the desulfated form of CCK8 nor human gastrin I or CCK4 release ACTH, and proposed CCK8 receptor antagonists such as proglumide and benzotript do not block CCK8's ability to evoke the secretion of ACTH. This pharmacological profile is distinct from CCK8 receptors in the brain or pancreas, suggesting that anterior pituitary CCK8 receptors may be a subclass of receptors for this peptide. Interestingly, CCK8 and CRF are colocalized in some of the same neurons of the parvocellular regions of the hypothalamic paraventricular nucleus that innervate the median eminence (Mezey et al., 1985). The levels of CCK8 in these neurons are regulated differently than in other central CCK8-containing neurons, suggesting a specific functional role of these cells. The physiologic role of CRF/CCK8 colocalization is, however, not established.

III. Intracellular Mechanisms of ACTH Release

A. Role of cAMP

CRF produces a variety of biochemical events within the corticotroph in stimulating ACTH release. cAMP could be a second messenger in the ACTH release response to CRF, as it may be for the secretion of many anterior pituitary hormones (Fig. 2). CRF activates adenylate cyclase in anterior pituitary membranes and raises cAMP levels both in primary cultures of the adenohypophysis and AtT-20 cells (Aguilera et al., 1983; Heisler et al., 1982; Heisler and Reisine, 1984). Catecholamines and VIP also elevate cAMP levels in AtT-20 cells (Reisine et al., 1982, 1983). Furthermore, forskolin, a diterpene that bypasses hormone receptors to stimulate adenylate cyclase, and cholera toxin are effective ACTH secretagogues (Heisler et al., 1982; Heisler and Reisine, 1984). The rise in intracellular cAMP leads to activation of a cAMP dependent protein kinase (Miyazaki et al., 1984; Litvin et al., 1984) which in turn catalyzes the phosphorylation of at least 10 distinct cellular proteins (Rougon et al., 1985). These phosphoproteins are localized to different subcellular fractions (nuclear, cytoplasmic, and cell membrane) which suggest a diversity of cellular actions of cAMP.

Attempts to establish that biological responses such as ACTH release are mediated by the cAMP-dependent protein kinase have in general

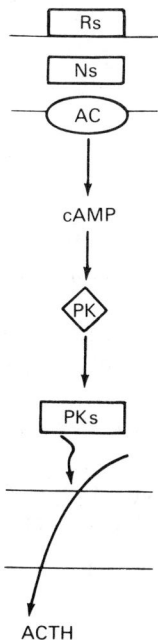

FIG. 2. cAMP regulates ACTH release. Stimulatory hormone receptors (Rs), such as those for CRF, VIP, and catecholamines, activate adenylate cyclase (Ac) via a guanine nucleotide stimulatory protein (Ns). This activation results in cAMP generation. cAMP activates a protein kinase (PK) which catalyzes the phosphorylation of cellular proteins (PKs). This phosphorylation event may lead to ACTH release.

proven difficult. Some investigators have pressure injected either the catalytic subunit of cAMP-dependent protein kinase or the inhibitor protein (PKI) of this enzyme into single neurons and examined the subsequent effect of this manipulation on the electrical activity of the cell either in response to cAMP or hormones (Nestler and Greengard, 1984). Such a procedure is inappropriate for studies on stimulus–secretion coupling since in general it is difficult to measure hormone secretion from a single cell which has been implanted with an electrode for injecting PKI. To circumvent this problem, a new technique was devised to deliver PKI into AtT-20 cells.

Liposomes have been used for several years as drug carriers *in vitro* and *in vivo* (Lesserman *et al.*, 1981). Although they do not spontaneously fuse or become incorporated into nonphagocytic cells in culture, they can be made to bind and be incorporated into specific cells by coupling covalently to the liposome surface a ligand, such as a monoclonal antibody, which recognizes some molecule on the surface of the target cells. This suggested that, if liposomes were made with encapsulated PKI and targeted to some determinant of the surface of AtT-20 cells, PKI could be released from the liposomes into the cell cytoplasm without severely altering the integrity of the corticotroph. The AtT-20 cells express the cell surface adhesion molecule N-CAM, and antibodies against this glycoprotein bind to these cells. Furthermore, the anti-N-CAM antibodies, once bound to AtT-20 cells, internalize. Treatment of the anti-N-CAM antibody-labeled AtT-20 cells with liposomes containing PKI and coupled to *Staphylococcus aureus* Protein A (a water soluble protein capable of binding to the Fc portion of several immunoglobulins, including rabbit IgG) totally abolished 8-bromo-cAMP and forskolin stimulation of ACTH release (Reisine *et al.*, 1986). The phosphorylation of cellular proteins following forskolin treatment was also prevented by this treatment, indicating that endogenous cAMP-dependent protein kinase activity was probably inhibited (Rougon *et al.*, 1985). Furthermore, the ACTH release response to CRF was not observable following incorporation of PKI into the AtT-20 cells (Reisine *et al.*, 1985a). These studies suggest an essential role of cAMP in mediating CRF-evoked ACTH secretion.

B. Role of Calcium

Calcium is necessary for the receptor-mediated release of ACTH from AtT-20 cells. Removal of calcium or application of calcium channel blockers prevents ACTH release induced by all secretagogues (Suprenant, 1982; Reisine *et al.*, 1982; Richardson, 1983). Recently, it was observed that 8-bromo-cAMP, forskolin, isoproterenol, and CRF increase cytosolic calcium levels as measured using the fluorescent compound Quin 2 (Guild *et al.*, 1986; Luini *et al.*, 1985) (Fig. 3). Voltage-sensitive calcium channels may be involved in the ACTH release response to CRF since this response is blocked by the calcium antagonists nifedipine and verapamil (Luini *et al.*, 1985). Furthermore, it was shown that 8-bromo-cAMP enhanced calcium conductance in whole-cell patch clamp preparations, suggesting that cAMP-dependent protein kinase may directly regulate calcium channels in AtT-20 cells (Luini *et al.*, 1985). This would be consistent with recent biochemical studies of Curtis and Caterall (1985) who showed that purified cAMP-dependent protein kinase applied to transverse tubular membranes phosphorylated a segment of the calcium channel labeled with a radioactive calcium channel antagonist. Interestingly, the PKI pretreatment blocked isoproterenol-evoked ACTH release from AtT-20 cells (Reisine *et al.*, 1985a). The β-adrenergic agonist has previously been shown to increase the frequency of action potentials in AtT-20 cells, and this electrical activity has been associated with an in-

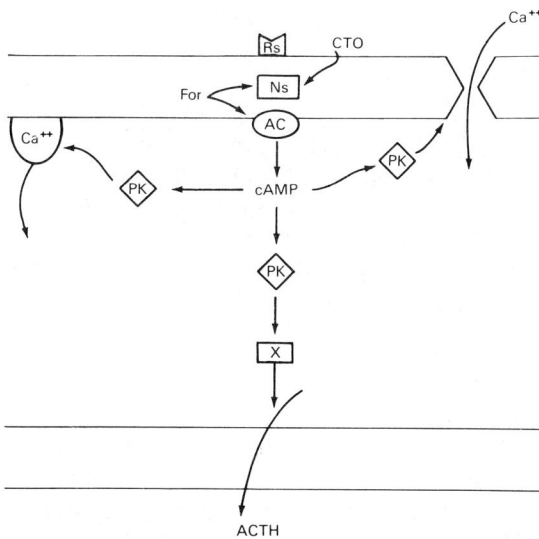

FIG. 3. cAMP regulates cytosolic calcium levels. Stimulation of adenylate cyclase either by hormone receptors, forskolin (FOR), or cholera toxin (CTO) leads to cAMP formation and protein kinase activation. cAMP-dependent protein kinase activation leads to a rise in cytosolic calcium levels (Ca^{2+}). This could be due to a phosphorylation of the Ca^{2+} channel or mobilization of Ca^{2+} from intracellular stores.

crease in Ca^{2+} conductance and ACTH release from AtT-20 cells (Suprenant, 1982). It was proposed by Suprenant (1982) that β-adrenergic receptor activation reduced the membrane potential necessary for action potential generation so as to trigger spike generation, calcium influx, and hormone secretion. The results with the PKI treatment would suggest that cAMP-dependent protein kinase is involved in activating that cellular mechanism controlling membrane potential.

In addition to regulating calcium channels, cAMP may also modify the actions of calcium in releasing ACTH. Using a superfusion apparatus to monitor the continuous release of ACTH, it was shown that forskolin potentiated the ACTH release response to K^+ but did not affect the ability of the membrane depolarizing agent to raise intracellular calcium levels (Guild et al., 1986). These results suggest that, rather than only raising cytosolic calcium levels, cAMP may facilitate the intracellular actions of calcium in triggering ACTH secretion. This interaction between cAMP and calcium may involve the attachment of ACTH secretory granules to the plasma membranes or other essential components of the ACTH release process.

The PKI treatment did not block the ACTH release response to K^+

(Reisine *et al.*, 1985a). This finding is not unexpected since K^+ does not activate cAMP-dependent protein kinase activity in corticotrophs (Litvin *et al.*, 1984). K^+ increases calcium influx (Richardson, 1983) and cytosolic calcium levels which may explain its ability to release ACTH (Guild *et al.*, 1986; Luini *et al.*, 1985). The mechanisms by which K^+ and cAMP regulate intracellular calcium must be different. Whether these findings imply multiple calcium channels in AtT-20 cells regulated differently or varying processes for triggering intracellular calcium translocation is not known. The results do indicate that there are multiple intracellular mechanisms involved in ACTH release.

C. Protein Kinase C

In addition to cAMP-dependent protein kinase, protein Kinase C is also present in AtT-20 cells, and phorbol esters, activators of this enzyme, evoke ACTH release from both rat anterior pituitary cells in culture and AtT-20 cells (Phillips and Tashijan, 1982; Phillips and Jaken, 1983) (Fig. 4). Phorbol esters induce the phosphorylation of multiple proteins in corticotrophs (Rougon *et al.*, 1985). Some of these phosphoproteins have a different size as those regulated by forskolin. Insertion of the PKI into AtT-20 cells did not prevent phorbol ester-stimulated ACTH release (Reisine *et al.*, 1985a). Furthermore, phorbol esters reduce cytosolic calcium levels in AtT-20 cells in contrast to cAMP (Luini *et al.*, 1985). These data suggest that activation of protein kinase C may induce a different cascade of events to release ACTH than cAMP-dependent protein kinase. The physiologic stimuli coupled to protein kinase C in AtT-20 cells is not known. It has been proposed that diacylglycerol (DAG) may serve as the endogenous stimulant of protein kinase C (Berridge, 1984; Nishizuka, 1983). DAG is formed in the conversion of phosphatidylinositol phosphates (PIPn) to inositol phosphates (IPn) by the enzyme phospholipase C. Recent studies have suggested an involvement of phosphatidylinositol (PI) turnover in ACTH release (Zatz and Reisine, 1985a).

D. Lithium and PI Turnover

Thus, lithium, an ion that blocks the phosphatase that catalyzes the breakdown of IPs to inositol, increases the levels of inositol mono- and bisphosphate in AtT-20 cells and evokes ACTH secretion from these cells as well as primary cultures of the rat anterior pituitary (Zatz and Reisine, 1985a). Elevated calcium levels also stimulate ACTH release and activate phospholipase C so that, in the presence of lithium, calcium can be ob-

7. ACTH SECRETION AND SYNTHESIS

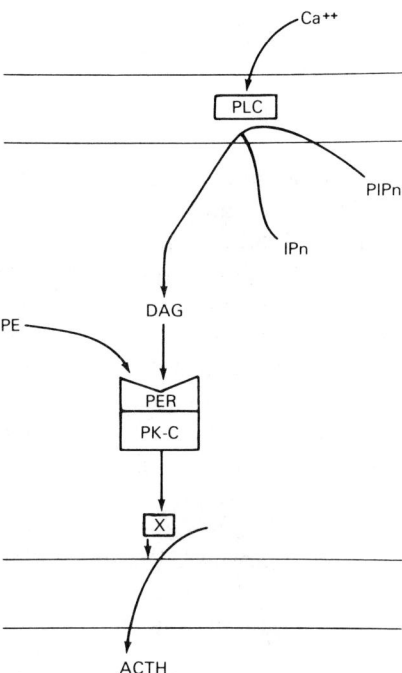

FIG. 4. Role of protein kinase C in ACTH release. Protein kinase C (PK-C) is present in AtT-20 cells, and activation of this enzyme by phorbol ester (PE) acting upon a phorbol ester receptor (PER) leads to ACTH release through an unknown mechanism (X). A possible endogenous activator of PK-C is diacylglycerol (DAG). DAG is a product of the breakdown of phosphatidylinositol phosphates (PIPn). This reaction is catalyzed by phospholipase C (PLC) which is activated by Ca^{2+}. Inositol phosphates (IPn) are also formed in this reaction.

served to increase inositol mono- and bisphosphate levels (Zatz and Reisine, 1985b). Whether lithium and extracellular calcium cause ACTH release by increasing inositol monphosphate, DAG, or some other compound in AtT-20 cells is not established. Chronic lithium exposure to AtT-20 cells diminishes phorbol ester-stimulated ACTH release (Zatz and Reisine, 1985a). If lithium increases DAG levels, that agent might be envisioned to alter protein kinase C or the phorbol ester binding site in such a way so as to desensitize the phorbol ester effect on ACTH release. Such a scheme might link the effect of physiologic stimuli on phospholipase C activity and hormone secretion in much the same manner as proposed for histamine release from mast cells (Nakamura and Ui, 1985). Recently, it was suggested that vasopressin may stimulate ACTH release from rat anterior pituitary cells in culture by increasing phosphatidic acid–phosphatidylinositol turnover, implying that this hormone may be a physiological stimuli of protein kinase C in corticotrophs (Raymond et al., 1985).

E. ARACHIDONIC ACID

Arachidonic acid and its metabolites have been shown to be involved in the receptor-mediated release of hormones such as growth hormone releasing factor, (Judd et al., 1985), prolactin (Camoratto and Grandison, 1985), or gonadotropin releasing hormone (Naor and Catt, 1985). It appears that this fatty acid may be involved in one of the intracellular events of ACTH secretion from AtT-20 cells (Luini and Axelrod, 1985). Mellitin, a peptide that activates phospholipase A_2, a calcium-dependent enzyme that liberates arachidonic acid from lipids, also stimulates ACTH secretion (Heisler et al., 1982), while the glucocorticoids, inhibitors of this enzyme, block the release of ACTH (Axelrod and Reisine, 1984). Arachidonic acid is metabolized by three pathways involving either cyclooxygenase (which forms prostaglandins and thromboxanes), lipoxygenase (which generates leukotrienes), or NADPH-dependent cytochrome P-450 (epoxygenase) enzymes (which yield epoxide metabolites). In determining which metabolites of the arachidonic acid-metabolizing enzymes are involved in ACTH secretion, inhibitors of these enzymes as well as phospholipase A_2 were added to AtT-20 cells (Luini and Axelrod, 1985). It was found that inhibitors of phospholipase A_2, lipoxygenase, and epoxygenase, but not cyclooxygenase, blocked the release of ACTH induced by CRF, isoproterenol, forskolin, and 8-bromo-cyclic AMP. These findings suggest, but do not prove, that metabolites of arachidonic acid formed via the epoxygenase and/or the lipoxygenase pathway are implicated in the stimulation of ACTH release induced by secretagogues.

IV. Inhibition of ACTH Release

A. GLUCOCORTICOIDS

Glucocorticoids consistently block basal and stimulated ACTH release from the anterior pituitary in the intact animal as well as cell preparations *in vitro* (Nakanishi et al., 1977; Birnberg et al., 1983; Civelli et al., 1983). Glucocorticoids may act through several mechanisms to inhibit ACTH secretion. Long-term treatment of animals or AtT-20 cells with dexamethasone reduces POMC mRNA levels, indicating an inhibition of ACTH synthesis at some pretranslational site (Nakanishi et al., 1977; Birnberg et al., 1983; Civelli et al., 1983). Short-term treatment (2 hr) of AtT-20 cells with dexamethasone seems to predominantly affect hormone-stimulated ACTH release rather than ACTH synthesis (Phillips and Tashijian, 1982). Treatment of AtT-20 cells with dexamethasone for short intervals, while inhibiting isoproterenol, CRF, and forskolin-stimulated ACTH release (Reisine et al., 1982) does not influence the activation of cyclic AMP-dependent protein kinase by these secretagogues (Miyazaki et al., 1984). The ability of these secretagogues to stimulate cyclic AMP accumulation was not affected by short- or long-term dexamethasone treatment (Reisine et al., 1982; Hook et al., 1982). Inability of glucocorticoids to block CRF-stimulated cyclic AMP formation was also reported in primary cultures of the anterior pituitary (Labrie et al., 1982; Giguiere et al., 1982). The precise manner by which glucocorticoids inhibit ACTH release is not known. The ability of glucocorticoids to antagonize the actions of a wide variety of secretagogues in releasing ACTH implies that they must inhibit an essential step in the hormone secretory process. Glucocorticoids inhibit phospholipase A_2 activity, and a possible mechanism for its actions is to reduce the generation of arachidonic acid, a product of phospholipase activity.

B. SOMATOSTATIN

Another hormone that reduces ACTH release from AtT-20 cells is somatostatin (SRIF) (Fig. 5). This 14-amino acid peptide is of hypothalamic origin and is known to block the secretion of growth hormone, prolactin, and thyroid stimulating hormone from the anterior pituitary (Brazeau et al., 1974; Vale et al., 1974). AtT-20 cells have SRIF receptors (Schonbrunn and Tashijian, 1976; Richardson and Schonbrunn, 1981) which when stimulated cause a reduction in ACTH secretion evoked by CRF, isoproterenol, VIP, cholera toxin or forskolin (Heisler et al., 1982). SRIF reduces the ability of these secretagogues to increase cyclic AMP accumulation (Heisler et al., 1982). This observation suggested that SRIF can block ACTH release by inhibiting the activation of adenylate cyclase. SRIF also inhibits forskolin-stimulated cyclic AMP formation in cyc⁻ variants of S49 lymphoma cells that are deficient in the guanine nucleotide stimulatory protein (N_s) required for most hormones to activate adenylate cyclase (Jakobs et al., 1983). From these data it was proposed that SRIF acted through a guanine nucleotide inhibitory protein (N_i) to reduce adenylate cyclase activity. A useful agent in studying the manner by which hormones inhibit adenylate cyclase activity is a toxin derived from the bacterium, *Bortedella pertussis*. Pertussis toxin induces the ADP-ribosylation of a 41,000 MW protein believed to be N_i (Katada and Ui, 1982). The toxin also blocks the inhibitory effects of hormones on adeny-

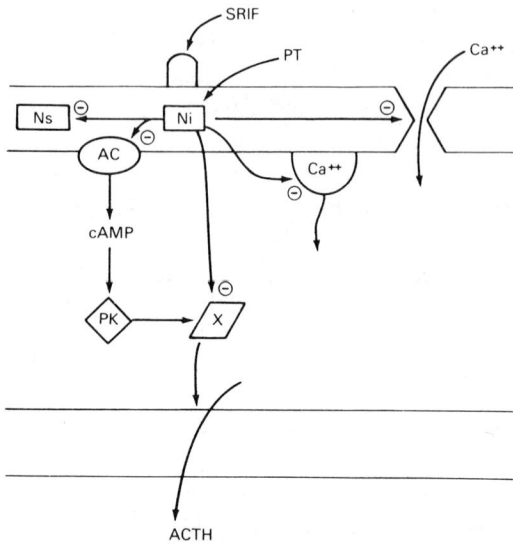

FIG. 5. Somatostatin (SRIF) inhibition of ACTH release. SRIF acts upon a membrane receptor to inhibit (−) adenylate cyclase. A guanine nucleotide inhibitory protein (Ni) mediates this inhibition. This protein is ADP-ribosylated in the presence of pertussis toxin (PT). In addition to blocking adenylate cyclase activation, SRIF reduces cytosolic Ca^{2+} levels by either inhibiting Ca^{2+} influx or mobilization. SRIF acts through an unknown mechanism to block 8-bromo-cAMP-directed ACTH release. Ns, guanine nucleotide stimulatory protein.

late cyclase in many tissues. The inhibition of growth hormone release by SRIF from primary cultures of the anterior pituitary is also blocked by pertussis toxin (Cronin et al., 1983). In membranes of AtT-20 cells, pertussis toxin induces the ADP-ribosylation of a 41,000 Da protein and also prevents the inhibitory effect of SRIF on forskolin, CRF, or isoproterenol-stimulated cyclic AMP formation and ACTH release (Reisine et al., 1985b). These findings suggest that SRIF can act through N_i to inhibit hormone induced ACTH release.

SRIF also blocks K^+, 8-bromo-cAMP-, and phorbol ester-stimulated ACTH release, suggesting a regulation of some nonadenylate cyclase intracellular mechanism by the peptide (Reisine et al., 1985b; Reisine, 1985). N_i may or may not mediate the effect of SRIF in blocking the ACTH release response to these secretagogues. Recent studies have suggested that N_i may couple hormone receptors to PI turnover, phospholipase C activity, and calcium mobilization (Nakamura and Ui, 1985; Bokoch and Gilman, 1984). Pertussis toxin does not abolish SRIF inhibition of K^+ or 8-bromo-cAMP-stimulated ACTH release (Reisine et al., 1985b; Reisine, 1985). Furthermore, SRIF does not inhibit K^+-stimulated calcium influx into AtT-20 cells (Richardson, 1983) or K^+- and 8-bromo-cAMP-induced calcium mobilization in these cells (Reisine and Guild, 1986). Thus, SRIF does not block the increase in intracellular calcium induced by ACTH secretagogues. However, SRIF does, by itself, lower cytosolic calcium levels, and this effect is blocked by pertussis toxin (Reisine and Guild, 1985). These results suggest that N_i may couple SRIF receptors to some mechanism, regulating basal intracellular calcium levels. This mechanism may be involved in the control of basal ACTH release which may be distinct from the regulation of evoked hormone secretion.

N_i does appear to regulate the sensitivity of SRIF receptors, even those involved in SRIF inhibition of K^+- or 8-bromo-cAMP-stimulated ACTH release. Thus, following pertussis toxin treatment, SRIF is less potent (100-fold reduction in potency) and less capable (50% loss of effectiveness) of inhibiting the ACTH release response to K^+ and 8-bromo-cAMP (Reisine, 1985). In addition, exposure of AtT-20 cells to pertussis toxin greatly diminishes the capability of SRIF and its analog to inhibit the binding of ^{125}I-labeled CGP 23996 (a highly potent and nonreducible SRIF analogue) to AtT-20 cell membranes (Reisine and Guild, 1985). By uncoupling N_i from SRIF receptors, pertussis toxin desensitizes SRIF receptors. These findings indicate that N_i is a crucial protein in regulating all of the inhibitory actions of SRIF.

V. Interactions of Corticotropin Releasing Factors

While the ACTH secretagogues can individually release ACTH, they may also act in concert to regulate the secretion of ACTH. Hypothalamic extracts are more potent in releasing ACTH than any secretagogue alone. These extracts appear to contain several different CRF-like factors. In primary cultures of the anterior pituitary, vasopressin added together with synthetic CRF induced a greater release of ACTH than the additive effect of the two peptides alone, indicating that vasopressin can potentiate the action of CRF (Giguere *et al.*, 1982; Vale *et al.*, 1983; Yates *et al.*, 1971; Gilles *et al.*, 1981; Rivier and Vale, 1983). While vasopressin does not alter cyclic AMP accumulation in the anterior pituitary, it causes a 4-fold potentiation in the stimulation of cyclic AMP synthesis by CRF, suggesting that vasopressin improves the efficiency of coupling between CRF receptors and adenylate cyclase (Giguere *et al.*, 1982). Epinephrine, by activating α_1-adrenoceptors in the anterior pituitary, also potentiates synthetic CRF stimulation of ACTH release and, like vasopressin, enhances the cyclic AMP response to CRF (Giguere and Labrie, 1983).

These findings indicate a synergism between vasopressin, α_1-adrenergic agonists and synthetic CRF in releasing ACTH. In AtT-20 cells, β_2-adrenoceptor agonists and CRF also interact to regulate ACTH release (Reisine et al., 1982). When CRF and isoproterenol are added together, the increase of ACTH secretion is less than additive, suggesting that these secretagogues act through a common mechanism. This intracellular mechanism is distal to cyclic AMP accumulation since the coapplication of CRF and isoproterenol produce additive effects on cyclic AMP formation. An intracellular site of interaction of these two secretagogues may be cAMP-dependent protein kinase since both agonists active this enzyme. VIP appears to release ACTH through a process independent of CRF or β-adrenergic agonists (Reisine et al., 1982). VIP, together with isoproterenol or CRF, causes an additive increase in both ACTH secretion and cyclic AMP production. These findings indicate that VIP may act on second messenger systems other than CRF or isoproterenol.

VI. Desensitization

A. CRF

While hormones can induce rapid and pronounced responses from cells, the presistent presence of the hormone can induce desensitization (Catt et al., 1979). Corticotrophs become refractory to CRF following prolonged exposure to this peptide (Reisine and Hoffman, 1983; Hoffman et al., 1985). This desensitization is manifest as a reduced maximal ability of CRF to stimulate both cyclic AMP formation and ACTH release. ACTH content is not grossly affected by CRF pretreatment, indicating that the cells are not depleted of the peptide hormone. Forskolin-stimulated cyclic AMP accumulation or ACTH release is not reduced by CRF pretreatment, suggesting that both adenylate cyclase and the intracellular mechanisms medicating stimulus-secretion coupling are unaffected in the desensitized cells. Thus, either CRF receptors are lowered in density or their coupling to adenylate cyclase is impaired. Studies using radiolabeled CRF to detect CRF receptors on anterior pituitary membranes indicate that desensitization involves the loss of CRF receptors (Wynn et al., 1983). Adrenalectomy, a procedure that abolishes the glucocorticoid feedback inhibition of CRF release in the hypothalamus, markedly decreased CRF receptor binding in the pituitary 4–6 days after surgery. The density of these sites returned almost to normal levels after treatment with dexamethasone. Thus, CRF receptors can be down-regulated, and this may explain the desensitization observed in primary cultures.

B. Vasopressin

Vasopressin responses are also desensitized on corticotrophs following prolonged vasopressin treatment (Antoni et al., 1985). Furthermore, vasopressin not only potentiates CRF-stimulated ACTH release, but it can also increase the ability of CRF to desensitize its own receptor (Hoffman et al., 1985). Pretreatment of primary cultures of the anterior pituitary with a fixed concentration of arginine-vasopressin and varying amounts of CRF reduces the amount of CRF needed to densensitize its receptors. Thus, vasopressin and CRF act synergistically to release ACTH and regulate CRF receptors.

C. β-Adrenergic Agonists

β-Adrenoceptors in many cell types are readily desensitized. This was also found to be the case for mouse pituitary tumor cells (Reisine and Heisler, 1983). Pretreatment of AtT-20 cells with isoproterenol results in a marked reduction of cyclic AMP formation and release of ACTH after restimulation with the catecholamines. Following pretreatment with isoproterenol there is a decreased binding of [^{3}H]dihydroalprenolol which becomes apparent after 20 hr of treatment. The reduced binding is associated with a decreased density of β-adrenoceptors but no change in receptor ligand affinity. The desensitization of the cyclic AMP accumulation and ACTH secretion responses were observed before there was a decrease in receptor density. These findings suggest that the desensitization of the β-adrenoceptor is a two-step process. The first is rapid in onset and shows a reduced capacity of catecholamines to elevate cyclic AMP accumulation and ACTH release. The second step is slower and is associated with a loss of β-adrenoceptors from cell membranes (down-regulation).

The rapid desensitization of the β-adrenoceptors to stimulation of cyclic AMP synthesis and ACTH release without changes in receptor density could be due to the uncoupling of the receptor from the adenylate cyclase complex, decreased activity of adenylate cyclase, or changes in the ACTH secretory process. These possibilities were examined by first pretreating AtT-20 cells with isoproterenol to reduce their responsiveness to cyclic AMP elevation and ACTH secretion by about 50%. The cells were then treated with forskolin to directly stimulate adenylate cyclase. The generation of cyclic AMP and release of ACTH in forskolin-treated cells were the same as those of the fully sensitized cells. This experiment indicates that, during the early densitization of the β-adrenoceptors, the adenylate cyclase and ACTH secretory mechanisms are normal and that the desensitization is due to an uncoupling of the receptor from adenylate cyclase.

β-Adrenoceptor desensitization is specific since CRF- and VIP-stimulated cyclic AMP accumulation and ACTH release were unaffected by catecholamine pretreatment of AtT-20 cells. CRF receptors on normal corticotrophs are also regulated independently of catecholamine receptors. The independent nature of the desensitization of these receptors indicates that the mechanisms involved in this process are specific for each receptor. Such a property would allow corticotrophs to respond to some stimuli despite the loss of responsiveness to other CRF-like substances.

A rapid desensitization of β-adrenergic receptors on normal corticotrophs may explain the apparent lack of ability of β-adrenergic agonists to stimulate ACTH release from primary cultures of the rat anterior pituitary. In fact, pretreatment of primary cultures of the anterior pituitary for 1 min with isoproterenol totally abolishes the subsequent ability of the agonists to increase intracellular cAMP levels (T. Reisine, unpublished results). The different time courses of ACTH secretagogues to induce homologous receptor densitization may explain their variable *in vivo* effects on ACTH release.

D. SOMATOSTATIN

Somatostatin can also regulate the sensitivity of its own receptor (Reisine and Axelrod, 1983; Reisine, 1984; Reisine and Takahashi, 1984). Preexposure of mouse anterior pituitary cells to SRIF lessens SRIF antagonism of CRF-, VIP-, isoproterenol-, and forskolin-stimulated cyclic AMP accumulation and ACTH release. SRIF pretreatment increases the formation of cyclic AMP in response to forskolin in these cells. This increase is delayed in onset, slow to recover, and blocked by the protein synthesis inhibitor, cyclohexamide. This suggests that prolonged treatment of AtT-20 cells with SRIF desensitizes SRIF receptor and causes a compensatory sensitization of adenylate cyclase through a process requiring protein synthesis. In cultures of brain cells, prolonged application of SRIF produced an adaptive response which was manifested as a reduced ability of SRIF to stimulate cell firing activity (Delf and Dichter, 1983). Continued pretreatment of anterior pituitary cells with SRIF reduced the peptides subsequent ability to inhibit growth hormone and thyroid-stimulating hormone (TSH) release (Smith and Vale, 1980). These findings indicate that SRIF receptors on many cell types can be self-regulated.

Interestingly, SRIF pretreatment of AtT-20 cells did not reduce the sensitivity of the peptide to inhibit K^+- or 8-bromo-cAMP-stimulated ACTH further, implying a different coupling of SRIF receptors to adenylate cyclase as compared to the effector systems mediating the actions of these other secretagogues on the ACTH release process (Reisine, 1984, 1985).

VII. The Multireceptor Release of ACTH *in Vivo*

The *in vitro* studies showing that ACTH release is stimulated by multiple factors raises the question of whether the release of ACTH *in vivo* is under multihormonal control. The injection of synthetic ovine CRF in rats causes an immediate rise in plasma ACTH levels (Rivier and Vale, 1982, 1983; Rivier *et al.*, 1982). This stimulation is dose dependent and neutralized by antibodies specifically raised against CRF. The ACTH release induced by an acute ether stress is partially blocked by CRF antibodies, indicating that molecules with similar immunologic characteristics, such as synthetic CRF, are stress mediators. The lack of total blockade of stress-evoked ACTH release in rats by the CRF antibodies suggests that hormones other than CRF are involved in promoting the release of ACTH *in vivo*.

Vasopressin also causes the release of ACTH in the intact rat. This stimulation is dose dependent and prevented by a vasopressin antagonist (Rivier and Vale, 1983; Knepel *et al.*, 1982). The physiological condition of the animal has an important role in the stimulation of ACTH secretion by vasopressin. Animals anesthetized with neuroleptics, opiates, and nembutal (conditions that block CRF release) respond to vasopressin with a smaller elevation in ACTH release than that found in awake, freely moving animals. Immunoneutralization of CRF in nonanesthetized rats also lowers stimulation of ACTH release by vasopressin, suggesting a dependence on CRF for the ACTH-releasing action of vasopressin. Vasopressin potentiates CRF-stimulated ACTH release in anesthetized rats, indicating that vasopressin and CRF act in a synergistic manner to regulate ACTH release *in vivo* as well as *in vitro*.

As described above, catecholamines increase the release of ACTH from primary cultures of the anterior pituitary and corticotroph tumor cells (Mains and Eipper, 1981; Tilders *et al.*, 1982; Berkenbosch *et al.*, 1981; Vale and Rivier, 1977). *In vivo* studies have also indicated that catecholamines can stimulate ACTH release by a direct action on the anterior pituitary (Mezey *et al.*, 1983). Peripheral injections of epinephrine increase plasma levels of ACTH in intact rats (Berkenbosch *et al.*, 1981; Tilders *et al.*, 1982). The rise in plasma ACTH levels caused by either epinephrine or (−) isoproterenol is stereospecifically blocked by propranolol, suggesting that β-adrenoceptors are linked to the *in vivo* release of ACTH. β-adrenoceptor agonists have been proposed to stimulate ACTH release *in vivo* by acting through "central mechanisms" since hypothalamic lesions of female rats prevents the rise in plasma ACTH levels induced by (−) isoproterenol (Vermes *et al.*, 1981). The process by which catecholamines gain access to the brain to initiate these central effects is not known. Previous work has demonstrated that catecholamines only minimally cross the blood–brain barrier (Weil-Malherbe *et al.*, 1959, 1961). In contrast to these studies, Mezey *et al.* (1983) found that (−) isoproterenol stimulated ACTH release from male rats in which the hypothalamus was separated from the pituitary by either stalk-transection or median eminence lesions. The effect of isoproterenol was blocked by propranolol but not by the selective β_1-adrenoreceptor antagonist, practolol. Salmefamol, a β_2-adrenoceptor agonist, also stimulated ACTH release in stalk-sectioned animals, indicating that β_2-adrenoceptors can mediate the *in vivo* stimulation of ACTH release by catecholamines. Isoproterenol-stimulated ACTH release in stalk-transected animals is blocked by dexamethasone pretreatment, suggesting that the ACTH release induced by β-adrenoceptor agonists originates from the anterior pituitary. These findings are consistent with the previous studies of Fortier (1951) and McDermott *et al.* (1950) who used pituitary transplants to examine the direct action of epinephrine on an ACTH-mediated response. In these studies, the anterior pituitary was placed into the anterior chamber of the eye of hypophysectomized rats. Injection of small quantities of epinephrine into the eye reduced the level of circulating white blood cells (eosinopenia), which is believed to accompany an increase of ACTH release. Similar injection of epinephrine into the other eye did not produce this response. These data as well as the findings in stalk-transected animals indicate that epinephrine can act directly on the anterior pituitary, possibly via β-adrenoceptors to stimulate ACTH release.

The relative importance of peripheral catecholamines as compared to CRF in mediating the effect of stress stimuli on ACTH release is not established. However, recent studies have shown that some forms of stress which may not be primarily mediated by CRF or central factors could instead be predominantly expressed through the actions of peripherally circulating epinephrine. Thus, insulin-induced stress stimulates ACTH release despite pituitary stalk transection and is blocked by propranolol (Mezey *et al.*, 1984). This stress is one of the most effective means to raise plasma epinephrine levels. The hypoglycemia induced by insulin may affect central neurons that in turn regulate the splanchnic nerve. This peripheral fiber system controls the release of epinephrine from the adrenal medulla.

Most forms of stress raise plasma catecholamine levels. However, these stressors also activate the hypothalamic–pituitary axis. The central control of ACTH release appears to be much more pronounced than the contribution provided by circulating catecholamines. Thus, the effect of a variety of stressors on ACTH secretion is relatively resistent to manipulations that antagonize the effects of peripheral catecholamines and is greatly attenuated by stalk section. In contrast, forms of stress that do not activate the hypothalamic–pituitary axis may act through this peripheral control system. Further studies will be necessary to clarify the role of central vs peripheral mechanisms in the control of normal forms of stress encountered by humans rather than the excessive number of, and possibly inappropriate, animal models presently employed.

VIII. Regulation of ACTH Synthesis

ACTH is derived from a larger prohormone, proopiomelanocortin (POMC) (Mains and Eipper, 1976). Within the last few years the sequence of the POMC gene has been elucidated and some of the enzymes involved in the processing of ACTH were characterized. This has allowed for a detailed analysis of the molecular mechanisms controlling ACTH synthesis.

The first known regulators of ACTH synthesis were the glucocorticoids. These steriods are manufactured in the adrenal cortex and ACTH is a potent stimulator of their production. The glucocorticoids induce many biological effects, including the feedback inhibition of ACTH synthesis (Harris, 1948). Glucocorticoids regulate ACTH synthesis through intracellular receptors that transport the steroids to the cell nucleus. The levels of POMC mRNA are lowered by glucocorticoids (Nakanishi et al., 1977), and these agents have been suggested to inhibit the rate of transcription of the POMC gene (Birnberg et al., 1983; Eberwine and Roberts, 1984). Whether, however, glucocorticoids and their receptors can bind to the POMC gene or some regulatory region controlling POMC gene expression has not been established.

The POMC gene in corticotrophs can also be stimulated by hormones. Infusion of CRF for 3 days into rats increases POMC mRNA levels in the anterior pituitary (Bruhn et al., 1984) (Fig. 6). Application of CRF onto AtT-20 cells for 4 hr significantly elevates the levels of POMC mRNA (Affolter and Reisine, 1985). This increase appears to be due to an activation of the POMC gene since the levels of a nuclear RNA species are larger than mature POMC mRNA, and having the expected size of the primary transcript of the POMC gene, were greater following CRF treatment. 8-Bromo-cAMP also increases the levels of POMC mRNA in AtT-

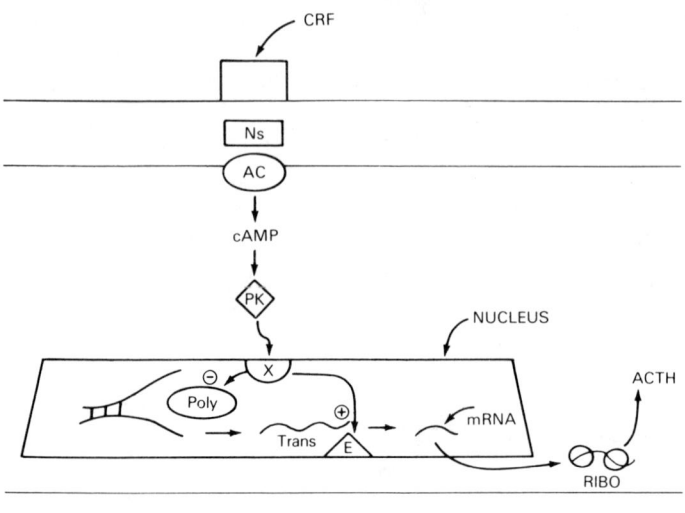

FIG. 6. CRF stimulates POMC gene expression. CRF activates adenylate cyclase (AC) and cAMP-dependent protein kinase (PK) to stimulate the POMC gene. This stimulation could be due to an enhancement of the affinity of the promotor region of the POMC gene for RNA polymerase (Poly), a higher transcription rate, or a change in the processing (E) of the primary transcript (TRANS) to mature POMC RNA. Once processed, the mature POMC mRNA leaves the nucleus and translation to form POMC occurs at the ribosomes (RIBO).

20 cells, suggesting that CRF may act through cAMP-dependent protein kinase to activate the POMC gene. This was tested by incorporating PKI into AtT-20 cells using liposomes. This manipulation prevented the increase in POMC mRNA levels seen with CRF or 8-bromo-cAMP treatment (Reisine *et al.*, 1985a). Furthermore, PKI treatment lowered basal POMC mRNA levels, further indicating a role for cAMP in controlling of the POMC gene.

CRF may regulate the POMC gene by inducing the phosphorylation of nuclear proteins through a cAMP-dependent mechanism. In another tumor cell line of the anterior pituitary, GH_3, it was observed that thyroid stimulating hormone releasing factor and cAMP increase prolactin gene transcription and phosphorylation of nuclear substrates (Murdoch *et al.*, 1982, 1983). In primary cultures of the anterior pituitary, growth hormone releasing factor (GHRF) and forskolin cause the phosphorylation of histone proteins in the nucleus as well as activate the growth hormone gene (Barinaga *et al.*, 1985). In AtT-20 cells, forskolin also stimulates the phosphorylation of nuclear as well as other cellular proteins (Rougon *et al.*, 1985). The manner by which these phosphorylation events enhance gene expression is not known. cAMP regulated phosphoproteins could induce promotor affinity changes, higher transcription rates, or other regulatory events occurring downstream from the start site of transcription.

Phorbol esters also increase POMC mRNA levels, and this effect is not blocked by PKI (Affolter and Reisine, 1985; Reisine *et al.*, 1985a). That two different protein kinases could regulate the same biologic event is not unprecedented. Both cAMP-dependent protein kinase and protein kinase C affect tyrosine hydroxylase activity in the striatum and catalyze the phosphorylation of the same serine residue on this enzyme (Albert *et al.*, 1984). In AtT-20 cells, phorbol esters may increase POMC mRNA levels through similar or different phosphorylation events as cAMP-dependent protein kinase. Interestingly, in primary cultures of the anterior pituitary, cAMP, but not phorbol esters, increases growth hormone gene transcription (Barinaga *et al.*, 1985). The presence of phosphoproteins regulated by one or the other of these protein kinases in different cell types of the anterior pituitary may be a major factor determining the mode of gene regulation in these cells by hormones and second messengers.

IX. Conclusion

The diversity of second messenger systems in corticotrophs presents a wide assortment of mechanisms for different extracellular releasing factors to evoke ACTH secretion. This diversity may act as a sort of "fail safe" system to ensure the continued responsiveness of the corticotroph to physiological stimuli despite conditions in which simpler systems would be expected to become desensitized. In fact, the corticotroph can become refractory to individual releasing factors, such as CRF, vasopressin, or catecholamines, but the desensitization is homologous and the cells remain responsive to other stress modulators (Axelrod and Reisine, 1984). Furthermore, the steady-state level of ACTH synthesis and release is determined by the resultant effect of stimulatory factors such as CRF and inhibitory hormones such as glucocorticoids on the corticotroph. Since the response to stress is essential for an animal's survival, the maintenance of responsiveness of ACTH-secreting cells through the multiplicity of extracellular stimuli and the diversity of intracellular second messengers may represent an inherent molecular characteristic to meet this goal.

References

Affolter, H. U., and Reisine, T. (1985). *J. Biol. Chem.* **260,** 15477–15481.
Aguilera, G., Harwood, J., Wilson, J., Morell, J., Brown, J., and Catt, K. (1983). *J. Biol. Chem.* **258,** 8039–8045.
Albert, K., Helmer-Matyjek, E., Nairn, A., Muller, T., Haycock, J., Greene, L., Goldstein, M., and Greengard, P. (1984). *Proc. Natl. Acad. Sci. U.S.A.* **81,** 7713–7717.
Antoni, F., Holmes, M., and Kiss, J. (1985). *Endocrinology* **117,** 1293–1299.
Axelrod, J., and Reisine, T. (1984). *Science* **224,** 452–459.
Barinaga, M., Bilezekijian, L., Vale, W., Rosenfeld, M., and Evans, R. (1985). *Nature (London)* **314,** 279–281.
Berkenbosch, F., Vernnes, I., Binnekade, R., and Tilders, F. (1981). *Life Sci.* **29,** 2249–2256.
Berridge, M. (1984). *Biochem. J.* **220,** 345–360.
Birnberg, N., Lissitzky, J., Hinman, M., and Herbert, E. (1983). *Proc. Natl. Acad. Sci. U.S.A.* **80,** 6982–6986.
Blumberg, P., Jaken, S., Konig, B., Sharkey, N., Leach, K., Jeng, A., and Yeh, E. (1984). *Biochem. Pharmacol.* **33,** 933–940.
Bokoch, G., and Gilman, A. (1984). *Cell* **39,** 301–308.
Brazeau, P., Rivier, J., Vale, W., and Guillemin, R. (1974). *Endocrinology* **94,** 184–197.
Bruhn, T., Sutton, R., Rivier, C., and Vale, W. (1984). *Neuroendocrinology* **39,** 170–175.
Camoretto, A. M., and Grandison, J. (1985). *Endocrinology* **116,** 1506–1513.
Catt, K., Harwood, J., Aguilera, G., and Dufau, M. (1979). *Nature (London)* **280,** 109–113.
Civelli, O., Birnberg, N., Comb, M., Douglass, J., Lissitzky, J., Uhler, M., and Herbert, E. (1983). *Peptides* **4,** 651–656.
Cronin, M., Rogol, A., Myers, C., and Hewlett, E. (1983). *Endocrinology* **113,** 209–215.
Curtis, B., and Catterall, W. (1985). *Proc. Natl. Acad. Sci. U.S.A.* **82,** 2528–2532.
Delf, J., and Dichter, M. (1983). *J. Neurosci.* **3,** 1176–1188.
Eberwine, J., and Roberts, J. (1984). *J. Biol. Chem.* **259,** 2166–2172.
Fortier, C. (1951). *J. Clin. Endocrinol.* **11,** 751–755.
Furchgott, H. (1972). *In* "Catecholamines" (H. Blaschko and E. Muscholl, eds.), pp. 283–285. Springer-Verlag, Berlin.
Giguere, V., and Labrie, F. (1982). *Endocrinology* **111,** 1752–1754.
Giguere, V., and Labrie, F. (1983). *Biochem. Biophys. Res. Commun.* **110,** 456–462.
Giguere, V., Cote, J., and Labrie, F. (1981). *Endocrinology* **109,** 757–763.
Giguere, V., Labrie, F., Cote, J., Coy, D., Suerras-Diaz, J., and Schally, A. (1982). *Proc. Natl. Acad. Sci. U.S.A.* **79,** 3466–3469.
Gilles, G., Linton, E., and Lowry, P. (1981). *Nature (London)* **299,** 355–356.
Guild, S., Itoh, Y., Kebabian, J., Luini, A., and Reisine, T. (1986). *Endocrinology* **118,** 268–279.
Harris, G. (1948). *Physiol. Rev.* **28,** 139–179.
Heisler, S., and Reisine, T. (1984). *J. Neurochem.* **42,** 1659–1666.
Heisler, S., Reisine, T., Hook, V., and Axelrod, J. (1982). *Proc. Natl. Acad. Sci. U.S.A.* **79,** 6502–6506.
Hoffman, A., Ceda, G., and Reisine, T. (1985). *J. Neurosci.* **5,** 234–241.
Hook, V., Heisler, S., Sabol, S., and Axelrod, J. (1982). *Biochem. Biophys. Res. Commun.* **106,** 1364–1371.
Jakobs, K., Aktories, K., and Schultz, G. (1983). *Nature (London)* **303,** 177–178.
Judd, A. M., Koike, K., and MacLeod, R. M. (1985). *Am. J. Physiol.* **248,** E438–E442.
Katada, T., and Ui, M. (1982). *Proc. Natl. Acad. Sci. U.S.A.* **79,** 3129–3135.

Kiss, J., Mezey, E., and Skirboll, L. (1984). *Proc. Natl. Acad. Sci. U.S.A.* **81,** 1854–1858.
Knepel, W., Benner, K., and Hertting, G. (1982). *Eur. J. Pharmacol.* **81,** 645–654.
Labrie, F., Veilleux, R., Lefebvre, G., Coy, D., Sueiras-Diaz, J., and Schally, A. (1982). *Science* **216,** 1007–1008.
Lesserman, L., Machy, P., and Barbet, J. (1981). *Nature (London)* **293,** 226–228.
Litvin, Y., Pasmantier, R., Fleischer, N., and Erlichman, J. (1984). *J. Biol. Chem.* **259,** 10296–10305.
Luini, A., and Axelrod, J. (1985). *Proc. Natl. Acad. Sci. U.S.A.* **82,** 1012–1014.
Luini, A., Lewis, D., Guild, S., Corda, D., and Axelrod, J. (1985). *Proc. Natl. Acad. Sci. U.S.A.* **82,** 8034–8038.
McDermott, W., Fry, E., Brobeck, J., and Long, C. (1950). *Yale J. Biol. Med.* **23,** 52–55.
Mains, R., and Eipper, B. (1976). *J. Biol. Chem.* **251,** 4115–4120.
Mains, R., and Eipper, B. (1981). *J. Cell Biol.* **89,** 21–28.
Mezey, E., Reisine, T., Palkovits, M., Brownstein, M., and Axelrod, J. (1983). *Proc. Natl. Acad. Sci. U.S.A.* **80,** 6728–6731.
Mezey, E., Reisine, T., Brownstein, M., Palkovits, M., and Axelrod, J. (1984). *Science* **226,** 1085–1087.
Mezey, E., Reisine, T., Skirboll, L., Beinfeld, M., and Kiss, J. (1985). *In* "Neuronal Cholecystokinin" (J. Vanderhaeghen and J. Crawley, eds.), pp. 152–156. New York Acad. Sci., New York.
Miyazaki, K., Reisine, T., and Kebabian, J. (1984). *Endocrinology* **115,** 1933–1945.
Murdoch, G., Rosenfeld, M., and Evans, R. (1982). *Science* **218,** 1315–1317.
Murdoch, G., Franco, R., Evans, R., and Rosenfeld, M. (1983). *J. Biol. Chem.* **258,** 15329–15335.
Nakamura, M., Nakanishi, S., Sueoka, S., Imura, H., and Numa, S. (1978). *Eur. J. Biochem.* **86,** 61–66.
Nakamura, T., and Ui, M. (1985). *J. Biol. Chem.* **260,** 3584–3593.
Nakanishi, S., Kita, T., Tau, S., Imura, H., and Numa, S. (1977). *Proc. Natl. Acad. Sci. U.S.A.* **74,** 3283–3286.
Naor, Z., and Catt, K. J. (1985). *J. Biol. Chem.* **256,** 222–229.
Nestler, E., and Greengard, P. (1984). *In* "Protein Phosphorylation in the Nervous System" (E. Nestler and P. Greengard, eds.). Wiley, New York.
Nishizuka, Y. (1983). *Trends Biochem. Sci.* **8,** 13–16.
Olivia, D., Nicosia, S., Spada, A., and Giannattasio, G. (1982). *Eur. J. Pharmacol.* **83,** 101–105.
Petrovic, S., McDonald, J., Snyder, G., and McCann, S. (1983). *Brain Res.* **261,** 249–259.
Phillips, M., and Jaken, S. (1983). *J. Biol. Chem.* **258,** 2875–2881.
Phillips, M., and Tashijian, A. (1982). *Endocrinology* **110,** 892–903.
Raymond, V., Leung, P., Veilleux, R., and Labrie, F. (1985). *FEBS Lett.* **182,** 196–200.
Reisine, T. (1984). *J. Pharmacol. Exp. Ther.* **229,** 14–20.
Reisine, T. (1985). *Endocrinology* **116,** 2259–2266.
Reisine, T., and Axelrod, J. (1983). *Endocrinology* **113,** 811–813.
Reisine, T., and Guild, S. (1985). *J. Pharmacol. Exp. Ther.* **235,** 551–557.
Reisine, T., and Heisler, S. (1983). *J. Pharmacol. Exp. Ther.* **227,** 107–114.
Reisine, T., and Hoffman, A. (1983). *Biochem. Biophys. Res. Commun.* **111,** 919–925.
Reisine, T., and Jensen, R. (1986). *J. Pharmacol. Exp. Ther.* **236,** 621–626.
Reisine, T., and Takahashi, J. (1984). *J. Neurosci.* **4,** 812–820.
Reisine, T., Heisler, S., Hook, V., and Axelrod, J. (1982). *Biochem. Biophys. Res. Commun.* **108,** 1251–1257.
Reisine, T., Heisler, S., Hook, V., and Axelrod, J. (1983). *J. Neurosci.* **3,** 725–732.

Reisine, T., Rougon, G., Barbet, J., and Affolter, H. U. (1985a). *Proc. Natl. Acad. Sci. U.S.A.* **82,** 8261–8265.
Reisine, T., Zhang, Y., and Sekura, R. (1985b). *J. Pharmacol. Exp. Ther.* **232,** 275–282.
Reisine, T., Rougon, G., and Barbet, J. (1986). *J. Cell Biol.* **102,** 1630–1637.
Richardson, U. (1983). *Endocrinology* **113,** 62–68.
Richardson, U., and Schonbrunn, A. (1981). *Endocrinology* **108,** 281–290.
Rivier, C., and Vale, W. (1982). *Science* **218,** 377–379.
Rivier, C., and Vale, W. (1983). *Endocrinology* **113,** 939–942.
Rivier, C., Brownstein, M., Spiess, J., Rivier, J., and Vale, W. (1982). *Endocrinology* **110,** 272–278.
Roberts, J., Phillips, M., Rosa, P., and Herbert, E. (1978). *Biochemistry* **17,** 3609–3618.
Rotsztejn, W., Benoit, L., Besson, J., Berand, G., Bluet-Pajot, M., Kordon, C., Rosselin, G., and Duval, J. (1980). *Neuroendocrinology* **31,** 282–286.
Rougon, G., Barbet, J., Affolter, H-U., and Reisine, T. (1985). *Soc. Neurosci. Abstr.* **11,** 1093.
Sabol, S. (1980). *Arch. Biochem. Biophys.* **203,** 37–48.
Sawchencko, P., Swanson, L., and Vale, W. (1984). *Proc. Natl. Acad. Sci. U.S.A.* **81,** 1883–1887.
Schonbrunn, A., and Tashjian, A. (1976). *J. Biol. Chem.* **253,** 6473–6478.
Smith, M., and Vale, W. (1980). *Endocrinology* **106,** 261A.
Spiess, J., Rivier, J., Rivier, C., and Vale, W. (1981). *Proc. Natl. Acad. Sci. U.S.A.* **78,** 6517–6521.
Suprenant, A. (1982). *J. Cell Biol.* **95,** 559–566.
Tilders, F., Berkenbosch, F., and Smelik, P. (1982). *Endocrinology* **110,** 114–120.
Vale, W., and Rivier, C. (1977). *Fed. Proc., Fed. Am. Soc. Exp. Biol.* **36,** 2094–2099.
Vale, W., Rivier, C., Brazeau, P., and Guillemin, R. (1974). *Endocrinology* **95,** 968–977.
Vale, W., Spiess, J., Rivier, C., and Rivier, J. (1981). *Science* **213,** 1394–1397.
Vale, W., Vaughan, J., Smith, M., Yamamoto, G., Rivier, J., and Rivier, C. (1983). *Endocrinology* **113,** 1121–1131.
Vermes, I., Berkenbosch, F., Tilders, F., and Smelik, P. (1981). *Neurosci. Lett.* **27,** 89–93.
Weil-Malherbe, H., Axelrod, J., and Tomchick, R. (1959). *Science* **129,** 1226–1227.
Weil-Malherbe, H., Whitby, G., and Axelrod, J. (1961). *J. Neurochem.* **8,** 55–61.
Westendorf, J., Phillips, M., and Schonbrunn, A. (1983). *Endocrinology* **112,** 550–557.
Wynn, P., Aguilera, G., Morell, J., and Catt, K. (1983). *Biochem. Biophys. Res. Commun.* **110,** 602–608.
Yates, F., and Maran, J. (1974). *In* "Handbook of Physiology" (R. Greep and E. Astwood, eds.), pp. 367–404. American Physiological Society, Washington, D.C.
Yates, F., Russel, S., Dallman, M., Hedge, G., McCann, S., and Dhariwal, A. (1971). *Endocrinology* **88,** 3–11.
Zatz, M., and Reisine, T. (1985a). *Proc. Natl. Acad. Sci. U.S.A.* **82,** 1286–1290.
Zatz, M., and Reisine, T. (1985b). *Soc. Neurosci. Abstr.* **11,** 657.

Index

A

Acetylation
  melanocyte-stimulating activity of ACTH and, 22
  of POMC peptides, 73
ACTH
  desensitization of release
    β-adrenergic agonits, 187–188
    CRF, 186
    somatostatin, 188–189
    vasopressin, 187
  effect of pro-γ-MSHs on aldosterone secretion and, 139–140
  historical background, 59–62
  inhibition of release of
    glucocorticoids, 182–183
    somatostatin, 183–185
  intracellular mechanisms of release of
    arachidonic acid, 182
    calcium, 178–180
    cAMP, 176–178
    lithium and PI turnover, 180–182
    protein kinase C, 180
  multireceptor release of
    catecholamines, 174
    cholecystokinin, 176
    vasoactive intestinal peptide, 174–175
    vasopressin, 175
    in vivo, 189–191
  regulation of synthesis, 191–193
  site of action in steroidogenic pathway
    calcium ions and, 193–195
    cholesterol transport, 98–100
    cyclic AMP and, 100–101
    phospholipids and, 107–110
    phosphorylation and, 102–103
    protein kinase C and, 105–107
  protein synthesis and, 101–102
  role of subcellular components, 110–119
  structure-function relationships
    adrenal-stimulating activity, 8–21
    bioassay, 7–8
    melanocyte-stimulating activity, 21–22
  structure of, 2–5
  related peptides, 5–7
ACTH receptors
  in 3T3-L1 cells, 54–55
  in rat adipocytes, 51–54
  detection of
    biological activity of ^{125}I-labeled ACTH, 34–35
    historical survey, 32–33
    studies with [^{3}H]ACTH, 35–36
    synthesis of radioligand with full biological potency, 36–38
  in human adrenocortical cells
    adult, 48–51
    fetal, 48
  characterization in rat adrenocortical cells
    binding characteristics, 38–40
    correlation of binding with cAMP synthesis and steroidogenesis, 40–45
    role of calcium, 45–48
Actin, response to ACTH and, 112
Active site, of ACTH, 10, 14, 21–22
Adenylate cyclase, 183
  ACTH binding sites and, 33, 34, 49–51
Adipocytes, rat, ACTH receptors in, 51–54
Adrenal cortex
  functions of, 90
  production of steroids by
    energy, 97
    pathway, 92–94
    11β-hydroxylase, 97

17α-hydroxylase, 95–96
21-hydroxylase, 96–97
3β-hydroxysteroid dehydrogenase-Δ^{4,5}-keto-steroid isomerase, 95
C_{27} side-chain cleavage, 94–95
substrate, 90–92
Adrenal-stimulating activity, of ACTH
effects of substitutions
N-terminal, 13–14
in other positions, 20–21
in positions 6–9 and their vicinities, 14–20
minimum structure essential for activity, 8–11
structure required for full activity, 11–12
β-Andrenergic agonists, desensitization of ACTH release and, 187–188
Aldosterone
action of ACTH on biosynthesis of, 129
cholesterol uptake by zona glomerulosa cells, 131–132
control of early and late pathways, 132–133
receptors on zona glomerulosa cells, 129–130
role of cAMP, Ca^{2+}, eicosanoids and ANF and, 130–131
biosynthetic pathway, 128–129
control of early and late pathways, 132–133
secretion, control of, 127–128
β-LPH, β-MSH and β-endorphin, 136–138
α-MSH and, 134–136
PRO-γ-MSHs, 138–142
Amidation
of ACTH, activity and, 11–12
of POMC peptides, 71, 73
Amino acid(s), basic, processing of POMC and, 71, 74
D-Amino acids, N-terminal substitution in ACTH and, 13
Angiotensin II, aldosterone synthesis and, 130, 132–133
Arachidonic acid, ACTH release and, 182
Arginine-8 residue, of ACTH, replacement of, 15, 19
Arginine-17 and 18 residues, of ACTH, replacement of, 20
Atrial natriuretic factors, action of ACTH and, 131

AtT-20-Dlbv cells, study of ACTH synthesis and, 60, 75
Avoidance conditioning, effects of ACTH and related peptides, 152–154

B

Behavioral effects, opiate-like, of ACTH and related peptides, 151–152
Behavioral studies, of effects of ACTH and related peptides in humans
elderly subjects, 159–160
mentally retarded individuals, 158–159
normal volunteers, 157–158
Binding site, of ACTH, 11, 18
Bioassay, of ACTH, 7–8
Bovine, ACTH of, 2, 3, 4
Bovine serum albumin, ACTH binding by adipocytes and, 52, 53
Brain
developmental studies, organizational influences of neuropeptides on, 162–164
POMC mRNA in, 65
processing of POMC peptides in, 73–74

C

Calcium ions
ACTH binding and, 33, 45–48, 51, 104
ACTH release and, 178–180
action of ACTH and, 103–105, 108, 115, 118
aldosterone secretion and, 130–131
Calmodulin, action of ACTH and, 105
Carbon, radioactive, labeling of ACTH with, 35
Catecholamines, ACTH release and, 174, 185–186
in vivo, 189–190
Cholecystokinin, ACTH release and, 176
Cholesterol
adrenal steroid production and, 90–92
aldosterone synthesis from, 128–129
transport, action of ACTH and, 98–100, 111, 112, 113, 114
uptake by zona glomerulosa cells, 131–132
Cholesterol ester hydrolase, response to ACTH and, 115
CLIP, relationship to ACTH, 6

INDEX

Colchicine, response to ACTH and, 112–113
Corticosterone
 ACTH bioassay and, 7–8
 conversion to aldosterone, 129
Corticotropin releasing factor(s), 173, 178
 action *in vivo*, 189
 desensitization to, 186
 interactions of, 185–186
 synthesis of ACTH and, 191–192
 vasopressin and, 175
Cyclic AMP, 179
 action of ACTH and, 31–32, 36, 100–101, 118
 aldosterone secretion and, 130
 synthesis, binding of ACTH and, 40–45
 release of ACTH and, 176–178
Cytochalasin B, adrenal steroidogenesis and, 111–112, 113
Cytochrome P-450
 adrenal steroidogenesis and, 94–95, 96, 97, 113, 115, 117
 aldosterone synthesis and, 128, 129, 133
Cycloheximide, effects on adrenal steroidogenesis, 101, 104, 106, 114
Cytoplasm, response to ACTH and cholesterol ester hydrolase, 115
 new proteins, 116–118
 nucleus, 118–119
 plasma membrane, 118
 sterol carrier protein, 115–116
Cytoskeleton, action of ACTH and, 110–113

D

Deoxyribonuclease I, response to ACTH and, 112
Deoxyribonucleic acid, ACTH sequencing and, 2
Desentization, of ACTH release
 β-adrenergic agonists, 187–188
 CRF, 186
 somatostatin, 188–189
 vasopressin, 187
Developmental studies, organizational influences of neuropeptides on brain, 162–164
Dexamethasone
 ACTH release and, 182–183
 regulation of POMC mRNA in anterior pituitary, 66–68

Diacylglycerol, protein kinase C and, 105, 106, 107, 108, 180
Dogfish, ACTH of, 2, 3

E

Eicosanoids, aldosterone secretion and, 131
Elderly subjects, behavioral effects of ACTH and related peptides on, 159–160
Electrophysiological effects, of ACTH and related peptides
 in animals, 160
 in human subjects, 160–162
Endogenous levels, of ACTH and related peptides, 164–165
Edorphins, in pituitary and brain, 73–74
β-Endorphin, 64
 ACTH binding and, 38
 effect on aldosterone secretion, 137
 hypoaldosteronism and, 138
 relationship to β-LPH, 6, 61
Energy, adrenal steroidogenesis and, 97
Ergocryptine, POMC mRNA levels in neurointermediate pituitary, 68–69

G

Gene, of POMC, structure in different species, 62–64
Gene transfer systems, identification of POMC processing enzymes and, 74–77
Glucocorticoids
 inhibition of ACTH release and, 182–183
 synthesis of ACTH and, 191
Glycine-10 residue, of ACTH, replacement of, 18–20
Glycosylation, of POMC peptides, 71, 73
Grooming, effects of ACTH and related peptides, 149–150
Guanine nucleotide binding, human adrenal glands and, 51

H

Haloperidol, POMC mRNA levels in neurointermediate lobe of pituitary, 68
Handling, early, developmental effects, 163
Histidine-6 residue, of ACTH, replacement of, 15

Horse, ACTH of, 2
Human ACTH of, 2, 3, 4
  behavioral studies on ACTH and related peptides
    elderly subjects, 159–160
    electrophysiological effects, 161–162
    mentally retarded individuals, 158–159
    normal volunteers, 157–158
11β-Hydroxylase, adrenal steroidogenesis and, 97
17α-Hydroxylase, adrenal steroidogenesis and, 95–96
21-Hydroxylase, adrenal steroidogenesis and, 96–97
3β-Hydroxysteroid dehydrogenase, $\Delta^{4,5}$-isomerase
  activity, ACTH and, 48
  progesterone synthesis and, 95
Hypoaldosteronism
  β-LPH, β-MSH and β-endorphin in, 128
  pro-γ-MSHs and, 141–142
Hypothalamus, stretching-yawning syndrome and, 150

I

Insulin, ACTH binding and, 38
Iodination
  of ACTH analog, 36–37
  of ACTH, biological activity and, 34–35

L

Learning, effects of ACTH and related peptides
  avoidance conditioning, 152–154
  visual discrimination and reversal learning, 154–156
Leucine aminopeptidase, ACTH activity and, 13
Lipolysis, ACTH and, 51–53
Liposomes, introduction of protein kinase inhibitor into pituitary cells and, 177–178
Lithium ions, ACTH release and, 180–181
Low density lipoproteins, response to ACTH and, 118
β-LPH, 60–61
  effects on aldosterone secretion, 136–137
  hypoaldosteronism and, 138
  mechanism of action, 137–138
  relationship to β-MSH, 6

Lysine-11 residue, of ACTH, replacement of, 18
Lysine-15 and 16 residues
  of ACTH, replacement of, 20–21

M

Melanocyte-stimulating activity, of ACTH, effects of substitutions, 21–22
Memory, effects of ACTH and related peptides, 153–154
Mental retarded individuals, behavioral effects of ACTH and related peptides on, 158–159
Methionine-4 residue, of ACTH, oxidation of, 20, 22, 34
Mitochondria, response to ACTH and 113–115, 117
Mouse, ACTH of, 2, 3, 4 MSH, active site of, 10
α-MSH
  cyclic analog, activity of, 5
  effects on aldosterone secretion, 134–135
  mechanism of action, 135–136
  relationship to ACTH, 5, 61
β-MSH
  effect on aldosterone secretion, 136, 137
  hypoaldosteronism and, 138
  mechanism of action, 137–138
  relationship to ACTH, 5

N

Nucleus, response to ACTH and, 118–119

O

Ostrich, ACTH of, 2, 3

P

Pathway, of adrenal steroidogenesis, 92–94
  11β-hydroxylase, 97
  17α-hydroxylase, 95–96
  21-hydroxylase, 96–97
  3β-hydroxysteroid dehydrogenase-$\Delta^{4,5}$-keto-steroid isomerase, 95
  C_{27} side-chain cleavage, 94–95
Peptides, ACTH-related, structure of, 5–7
Pertussis toxin, ACTH release and, 183–185

Phe^2, Nle^4-ACTH (1-38)
  iodination of, 36-37
  synthesis and biological potency of, 36
Phenylalanine-7 residue, of ACTH, replacement of, 15, 22
Phosphatidylinositides
  ACTH release and, 180-181
  action of ACTH and, 107, 108, 109
Phospholipase C, ACTH release and, 180-181
Phospholipids, action of ACTH and, 107-110
Phosphorylation
  action of ACTH and, 102-103, 115
  cAMP and, 176
  protein kinase C and, 180
  synthesis of ACTH and, 192-193
  of ACTH, 4
Pig, ACTH of, 2, 3, 4
Pituitary anterior lobe
  POMC gene regulation in, 66-68
  processing of POMC in, 71-73
  factors affecting of POMC peptides from, 66
  neurointermediate lobe
  POMC gene regulation in, 68-69
  processing of POMC in, 71-73
Pituitary factors, non-ACTH, controlling aldosterone secretion, 133-134
  β-LPH, β-MSH and β-endorphin, 136-138
  α-MSH, 134-136
  PRO-γ-MSHs, 138-142
Plasma membrane, response to ACTH and, 118
Polyamines, stretching-yawning syndrome and, 150
Polylysine, ACTH binding and, 35-36, 38
Potassium ions, ACTH release and, 179-180
Pregnenolone, adrenal steroid synthesis and, 93, 99
Proenkephalin, processing by transformed cells, 75-77
  vaccinia virus and, 77-80
PRO-γ-MSHs, 138-139
  effects on aldosterone secretion, 139-140
  hypoaldosteronism and, 141-142
  mechanism of action, 140-141
Pro-opiomelanocortin
  approaches to identification of processing enzymes
    gene transfer systems, 74-77
    use of vaccinia virus as transformation vehicle, 77-80
  bioavailability, age and, 159
  distribution and site of synthesis of derived peptides, 64-66
  processing, 6-7
  in pituitary and brain, 70-71
  regulation of expression of genes
    in anterior lobe of pituitary, 66-68
    differential regulation, 70
    factors affecting peptide secretion from pituitary, 66
    in neurointermediate lobe of pituitary, 68-69
  relationship to ACTH, 6
  sequencing of, 2
  structure of gene and protein in different species, 62-64
Protein(s)
  new, response to ACTH and, 116-118
  miscellaneous proteins, 118
  rat adrenal, 117-118
  γ-1 cells, 117
  of POMC, structure in different species, 62-64
Protein kinase C
  ACTH release and, 180
  action of ACTH and, 105-107, 118
Protein synthesis, site of action of ACTH and, 101-102
Proteolysis, of POMC peptides, 72-73
Pseudogene, for POMC, 63
Puromycin, 101, 106

R

Rat, ACTH of, 2, 3, 4
Rat adrenal, response to ACTH, new proteins and, 117-118
Reversal learning and visual discrimination, effects of ACTH and related peptides, 154-156

S

Salmon, ACTH of, 3, 4
Serine-1 residue, of ACTH, replacement of, 22
Sheep, ACTH of, 2, 3, 4

C_{27} Side-chain cleavage, adrenal steroidogenesis and, 94–95, 99, 114
Signal sequence, removal from POMC, 70–71
Social behavior, effects of ACTH and related peptides, 150–151
Sodium depletion, aldosterone secretion and, 133
Somatostatin
  desensitization of ACTH release and, 188–189
  inhibition of ACTH release and, 183–185
Steroidogenesis
  ACTH binding and, 40–43, 50–51
  adrenal, stimulation of, 97–98
  calcium and, 46–47
Sterol carrier protein, response to ACTH and, 115–116
Stress
  cholesterol metabolism and, 92
  effects of ACTH and related peptides, 148–149
    grooming, 149–150
    social behavior, 150–151
    stretching-yawning syndrome, 150
  ACTH release and, 190–191
Stretching-yawning syndrome, effects of ACTH and related peptides, 150
Structure, of ACTH
  primary, 2–4
  secondary, 4–5
Substrate, for adrenal steroid production, 90–92

Synthesis, of ACTH, 5
N-terminal substitutions, ACTH activity and, 13–14
3T3-L1 cells, ACTH receptors in, 54–55

T

Tritium, labeling of ACTH with, 35
  binding sites and, 35–36
Tryptophan-9 residue, of ACTH, replacement of, 17–18, 22
Turkey, ACTH of, 2, 3

V

Vaccinia virus, use as transforming vehicle, identification of POMC processing enzymes and, 77–80
Vasoactive intestinal peptide, ACTH release and, 174–175, 186
Vasopressin, ACTH release and, 175, 185
  desensitization, 187
  in vivo, 189

W

Water, bound, cytoskeleton and, 111
Whale, ACTH of, 2, 3, 4

Z

Zona glomerulosa cells, ACTH receptors on, 129–130

RAYMOND H. FOGLER LIBRARY
DATE DUE

BOOKS ARE SUBJECT TO RECALL AFTER TWO WEEKS

~~T 1 5 1987~~